The Earth
and Its Satellite

The Earth and Its Satellite

Edited by John Guest

Rupert Hart-Davis London

Granada Publishing Limited
First published 1971 by Rupert Hart-Davis Ltd
3 Upper James Street London W1R 4BP

Copyright © 1971 by Rupert Hart-Davis Ltd
All rights reserved. No part of this publication
may be reproduced, stored in a retrieval system, or
transmitted, in any form or by any means, electronic,
mechanical, photocopying, recording or otherwise,
without the prior permission of the publishers.

ISBN 0 246 64051 0
Printed in Great Britain
by Jarrold & Sons Ltd, Norwich

Contents

Preface   7

Chapter 1   Historical Introduction   9

The Earth and the Moon in Ancient Times—Lunar Study Since the Renaissance—Origin of Lunar Features—The Study of the Earth Since the Renaissance

By J. B. Murray, B.A.

Chapter 2   Astronomy of the Earth-Moon System   17

The Orbital Paths of the Earth and Moon—The Moon's Orbit—The Inclination of the Moon's Orbit—Phases of the Moon and Earthshine—Eclipses of the Sun and Moon—The Rotation of the Moon about its Axis—Libration—The Shape of the Moon—The Origin of the Moon

By E. L. G. Bowell, B.Sc.

Chapter 3   The Interiors of the Earth and Moon   29

Introduction—Earthquakes—Internal Composition—Magnetic Fields—Internal Heat—Variations in Density—Conclusions

By R. Mason, B.A., Ph.D.

Chapter 4   Structure and Tectonics of the Earth   37

Mobile Belts and Stable Areas, Oceans and Continents—Mid-oceanic Ridges and Ocean-floor Spreading—Mountains—Island Arcs, Mobile Continental Margins, and Disappearing Ocean Floor—Transcurrent Faults and Transform Faults—A Theory of the Movements of the Earth's Crust—Plate Tectonics—The Stable Regions, and Isostasy

By R. Mason, B.A., Ph.D.

Chapter 5   Magmatic and Volcanic Processes on Earth   48

The Nature of Magma and Igneous Activity—The Rôle of Volcanic Gases—Chemical and Mineralogical Composition of Igneous Rocks—Basalts and Basaltic Volcanicity—Andesites: Volcanicity of Mountain Chains and Island Arcs—Granites and their Volcanic Equivalents—Craters and Calderas

By M. K. Wells, B.Sc., M.Sc., Ph.D., D.Sc.

Chapter 6   Surface Processes on Earth   66

Weathering—Erosion—Geological Work of Rivers—Underground Water—Geological Work of Ice—Geological Work of the Sea—Erosion in Arid Regions—The Net Result of Erosion on Earth—Sedimentation—Continental Deposits—Mixed Continental-marine Deposits—Marine Deposits—The Oceans

By A. J. Smith, B.Sc., Ph.D.

The Earth and Its Satellite

Chapter 7 A Satellite's View of the Earth — 88

Introduction—Uniqueness of Earth's Atmosphere—Surveillance Photography from Orbital Altitudes—A New Look at Cultivated Areas

By K. Fea, B.Sc.

Chapter 8 Surface of the Moon — 100

Earth-based Observations—Observation from Orbiting Space Vehicles—Lighting Conditions—The Colour of the Lunar Surface—Lunar Gravity—Mascons—The Lunar Atmosphere—Temperature—Ice Under the Surface—Permafrost—The Meteorite and Solar Wind Environment of the Moon

By E. L. G. Bowell, B.Sc.

Chapter 9 Geology of the Moon — 120

The Maria and Highlands—The Far Side—Lava Flows in the Maria—Mare Ridges—Rocks from the Maria—Volcanic Craters and Vents in the Maria—Small Cratering and the Regolith—Large Ray Craters—Craters in the Highlands

By J. E. Guest, B.Sc., M.Sc., Ph.D.

Chapter 10 Dating Rocks from Earth and Moon — 149

What is Meant by Radioactive Decay?—What is the Rate of Radioactive Decay?—Decay Schemes Useful to the Geologist—How to Date a Rock—What Rocks and Minerals can be Dated using the Potassium-argon Method?—The Rubidium-strontium Method—Uranium and Thorium-lead Methods—The Contribution of Geochronology to the Earth and Planetary Sciences—Age of the Earth—Meteorite Ages—The Age of the Moon—Conclusion

By M. R. Wilson, B.Sc., Ph.D.

Glossary — 157

Index — 161

Preface

In this era of space exploration the Moon and the planets have suddenly come well within the sphere of study of Earth scientists—geologists, geochemists, geophysicists and meteorologists, men who had previously given most of their attentions to the study of the planet on which we live, Earth. As data have come pouring back from the Moon during the past few years it has been possible to make considerable advances in the study of our satellite. Detailed photographs of the lunar surface can be analysed in much the same way as we analyse photographs taken of the Earth's surface from aircraft; and thus geologists are able to make geological maps of the Moon similar to those of Earth using the normal techniques of *photogeology*. Rock samples returned by American manned missions and Russian unmanned missions can be studied by geologists in exactly the same way as rocks collected on Earth. By setting up instruments on the lunar surface, the geophysicist is able to extend his terrestrial techniques to the Moon, examining the internal structure and the forces that operate on and near this body.

The space age started with the first successful launching of a Russian craft, Sputnik 1, in October 1957. Within a month of this another Sputnik was orbiting the Earth with the world's first 'astronaut'—a dog! In April 1961 the dog was replaced by Major Gagarin in his momentous flight lasting nearly two hours in the Russian craft Vostok 1. Just less than a year later American technologists had put Colonel Glenn into orbit.

By the end of 1966 the Russians and Americans had put thirty-seven astronauts into space, the longest flight, by Borman and Lovell in Gemini 7, lasting more than 330 hours. In order to accomplish the pledge of America's National Aeronautics and Space Administration to put a man on the Moon by 1970 it was necessary to do more than this; the surface of the Moon had to be examined carefully to ensure that a spacecraft could land safely, and also to look for good landing sites. The first step in this programme was to send probes to crash into the lunar surface, taking pictures as they approached the Moon until the moment of impact. Three of these Ranger probes were successful and sent back remarkable pictures of the surface details in 1964–5. The Russians had also sent Lunik and Zond probes to photograph the lunar far side that had never been seen before.

The next stage, having confirmed that the lunar surface presented a reasonable landing ground, was to attempt a soft landing with an unmanned craft. The Russian Luna 9 was the first to achieve this, followed within a few months by the American Surveyor 1 which sent back the most detailed pictures ever, with resolutions down to one millimetre. In all, five Surveyor craft were successfully landed, each carrying out more complex examinations of the surface, including making the first chemical analyses.

During this time five American Orbiter craft were launched to circle the Moon, taking thousands of photographs to give an almost Moon-wide photo coverage. Preliminary Apollo landing sites were chosen from these.

The story of the first few Apollo missions is too well known to millions of television watchers all over the world to be described in any detail here. Most readers will remember the first manned orbit of the Moon at Christmas 1968. During this flight and another one early in the following year many more photographs were taken; but this time, instead of being transmitted back to Earth by radio, the actual films were brought back by the astronauts. Nobody can have forgotten the Apollo 11 flight when man first stepped on to the Moon; by the end of 1970 two successful lunar landings had been accomplished.

Rocks brought back by the Apollo 11 and 12 missions were studied by more than 600 scientists from many countries; American scientists were in the majority, followed by those from Britain.

Although Russia has not at the time of writing attempted a lunar-manned landing she has made an important step forward in lunar and planetary studies by collecting rocks from the Moon with unmanned probes. During 1970 rocks were collected from a site in Mare Fecunditatis and brought back to Earth for study. With this facility, rocks could be collected from many sites over the Moon, not only at a lower cost but from areas where it may be unwise to land a man.

Clearly we are far from understanding the Moon in the same way we do the Earth, but we have gone a long way towards solving some of the fundamental problems and in formulating the problems yet to be solved; we can at least study the Moon on the same footing as we study the Earth.

It would be unfair to emphasise too strongly the lunar results without pointing out that the 'tables have been turned' on Earth scientists. With space-probes we are now able to look back at the Earth with the eye of an astronomer in a way that was once reserved only for the Moon and planets. This means that instead of looking at small parts of our world in detail we can now 'step back' and look at it on the global scale: examine major features in the Earth's crust previously unrecognised and carry out more extensive studies of the large-scale characteristics of our atmosphere, enabling a better understanding of meteorology. Thus we are in an even better position to compare the Earth with its fellow companions in the solar system.

During this past decade geologists and geophysicists have made a number of important advances in the study of the Earth. The most important of these is the formulation of a global theory for the Earth: this is the theory of *plate tectonics*, which provides us with a way of explaining the major phenomena of the Earth's crust: faulting, mountain building, volcanism and continental drift can now be explained by one theory in a manner that is convincing to most Earth scientists. Thus, for the first time in the history of our science, workers in many different fields and many parts of the world can all contribute to a single theory; the theory may well prove wrong, at least in part, but it will almost certainly lead us nearer to an understanding of our planet.

In planning this book it was impossible to cover all the relevant fields. However, we feel it necessary to apologise for two obvious omissions in the field of geology: palaeontology and economic ore deposits. Palaeontology, or the study of fossils, has played a most important part in understanding the history of the Earth and the formulation of many geological concepts. Indications are, however, that this subject will not play a part in lunar geology, as to the best of our knowledge there has been no life on the Moon. Reluctantly then we have not done justice to this subject here. It is also true that one of the most important uses of geology has been in prospecting and working ore deposits that are so important to our everyday life. Quite probably valuable minerals will be discovered on the Moon in the future.

Our aim in this work is to consider the Earth and Moon together, in an attempt to show how these two neighbours in space have developed and to explain why they are so different. The important thing to remember throughout the book is that the Earth is a planet with an atmosphere and water in a liquid state: in contrast the Moon has no atmosphere of any consequence. Thus on Earth we have a constantly active eroding force—that of running water. In only a short time the whole surface of the Earth can become modified by this process alone. As well as this the Earth is a mobile planet with internal forces changing the surface all the time: mountains are raised up, continents are moving and land is disappearing below the sea.

By comparison the Moon, apparently, is now almost dead. The only known eroding force is that of impacting meteorites and micro-meteorites; these bring about changes only very slowly. Added to this, the Moon's crust is now relatively inactive.

In its early history the Moon was a very active body and, as we shall see, large volumes of lava were erupted on its surface; but this activity appears to have ceased before the time for which we have a complete record in the rocks of what was happening on Earth. It is likely that the main features of the Moon's geography were established before life was created on the Earth, and that by studying the Moon we will extend knowledge of the history of the Earth-Moon system to near the time when these bodies were formed in the solar system.

John E. Guest, November 1970

Chapter One

Historical Introduction

J. B. MURRAY

The Earth and the Moon in Ancient Times

The Earth and its relation to other bodies in the universe has occupied man's thought from earliest times. In prehistoric days speculation was fanciful and mythical, but from the earliest written records we find knowledge about the Moon steadily growing to such an extent that men were able to infer and speculate with ever-increasing accuracy.

The observations and deductions of the Ancient Chaldeans and Egyptians are the earliest recorded. The pyramids, built around 4000 BC, are orientated to face the rising points of various astronomical bodies and are thus indicative of an advanced state of astronomical knowledge, as well as political organisation. In England the building of Stonehenge, begun about 2000 BC and modified in 1700 BC, indicates that Neolithic man of that period had accumulated a considerable knowledge of movements of the Sun and Moon. The stones and arches are aligned with the extreme positions of the rising and setting points of the Sun and Moon, and the 56 'Aubrey Holes' may well represent a seasonal eclipse cycle. The whole structure could thus be used to mark the phases of the Moon and to predict eclipses. Recent work has shown that a number, and perhaps all, of the stone circles of megalithic man were astronomical 'observatories' for marking the extreme positions of moonrise and thus keeping the calendar. Indications are that there was quite a flourishing school of astronomical teaching and knowledge in the England of 1700 BC, perhaps the result of one single megalithic genius, for the knowledge seems to have developed rapidly and been forgotten just as rapidly.

The main purpose of observing the Moon was undoubtedly timekeeping, apart from the more religious interest in predicting eclipses, for the Moon is an ideal timekeeper, it being necessary just to look at the phase, and not its position in the sky as with the Sun. Survivals of this timekeeping are universal; the *month* (a corruption of 'moonth') is the time from one new Moon to another, though now adjusted so that there are an exact number in each year, and the *week* is the period of one of the quarters of the Moon, though again this has been altered to an exact number of days.

What megalithic man actually thought about the physical existence of the Earth and the Moon we cannot tell. The Chaldeans and Egyptians did not get beyond thinking that the Earth was flat and the Moon a goddess. It was not until the comparatively freethinking climate of Ancient Greece that men began seriously to enquire into the causes of the phenomena they observed. Thales, in the seventh century BC, introduced Egyptian astronomy into Greece. This introduction kindled the imagination of many, in particular Pythagoras (his exact dates are unknown), whose remarkable theories represent the first efforts at intelligent interpretation. He taught that the Earth is a sphere, and that it rests without support at the centre of the universe. A fascinating aspect of this idea is that he seems not to have had any real evidence, but to have derived it by analogy with the Moon, which he knew from the observation of its phases to be a sphere and not the flat disk that one might deduce from the unchanging marks upon its surface.

A century later Philolaus introduced the idea that the Earth moves in space, together with the

Sun, Moon and planets, round some central fire, an idea that did not arise again until after the time of Herschel.

Aristarchus of Samos (310–230 BC) represents another landmark in astronomical knowledge. He hit on the correct solution that the spherical Earth rotates on its axis and moves round the Sun, while the Moon revolves about the Earth. This idea was unfortunately rejected, probably owing to its being so far removed from common sense! Aristarchus was also the first to measure the relative distances of the Moon and the Sun by an ingenious geometrical method. He realised that if the Sun was at infinity and the Moon near the Earth, the Moon would appear half illuminated when at exactly 90° from the Sun. By measuring its actual angle from the Sun when half illuminated he found it to be less than a right angle, and from the actual figure deduced that the distance of the Sun was about 18 to 20 times the distance of the Moon. This was considerably in error, the real value being about 400 times the distance of the Moon, but this was not surprising owing to the extreme difficulty in measuring exactly half-phase. The main point was that he could demonstrate that the Sun was considerably farther away than the Moon. He also realised that the shadow crossing the Moon during a lunar eclipse was the shadow of the Earth, and that its curved shape provided a further proof that the Earth was spherical. From observations of the curvature of this shadow he deduced that the Earth was about three times the size of the Moon—a result very close to the actual value of 3·67 times.

Aristotle (384–322 BC) had provided further proof that the Earth is spherical by noting that as one moved north and south the stars and other celestial bodies changed their positions, and even disappeared or appeared in relation to the horizon. Eratosthenes (276–195 BC) later used this observation to make the first determination of the size of the Earth. He found by measuring the shadow from a building of known height that the Sun at midday was 7° from the vertical at Alexandria, whereas at the same time it was known to be vertical at Aswan in Upper Egypt. He thus deduced that the distance from Aswan to Alexandria (5,000 *stadia*) was about one-fiftieth the circumference of the Earth, the circumference thus being about 250,000 *stadia*. If we give him the benefit of the doubt as to which particular value of the *stadium* he was working with, his value gives a diameter of the Earth only about 80 km greater than the true value. He also made an estimate of the tilt of the Earth's axis from observations of the Sun at the solstices, the value he obtained, 23° 51′, being only 7′ in error.

Original Greek astronomy virtually ceased after the second century BC, and no significant advances were made again until the Middle Ages. However, one sideline of scientific knowledge is the speculation of science-fiction writers, and it is interesting to note that one ancient author's dreams of interplanetary travel have come down to us. Lucian of Samos (second century AD) wrote an account of an adventurer who was taken to the Moon by a waterspout which sucked him and his ship skywards while sailing beyond the Pillars of Hercules. In another book he journeyed to the Moon by making a pair of wings, launching himself from Mount Olympus and flying there.

Lunar Study Since the Renaissance

The next major advances in the study of the Moon, apart from improved observations and theory of its motion by Tycho Brahe (1546–1601) and Johannes Kepler (1571–1630), followed the invention of the telescope in about 1608 by the Dutchman Hans Lippersheim. The first known telescopic observations of the Moon were made in July 1609 by an English mathematician named Thomas Harriot (1560–1621) and his pupils and friends. He made accurate drawings and from these constructed the

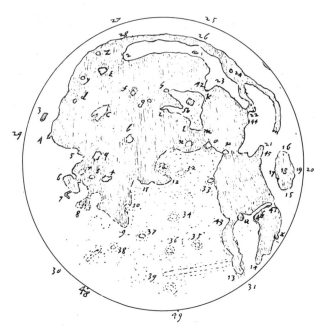

Plate 1   Harriot's map of the Moon (1610).

first map of the Moon, upon which a number of features may be recognised (*Plate* 1). Despite the energetic requests of his friends, Harriot never published his observations. Galileo (1554–1642) began telescopic observing late in 1609, and though his drawings are not as accurate as those of Harriot he quickly published them together with an account of the lunar surface features, contradicting the accepted ideas of Aristotle that the Moon was a perfect sphere, with no irregularities or colour differences to stain that perfection. This caused much controversy, and some authorities suggested that the irregularities were not the real surface but were overlain by a layer of invisible crystal with a perfectly smooth surface, and thus that Galileo's observations did not invalidate Aristotle's theory! Galileo also discovered that the Moon, while keeping the same face towards the Earth, does in fact slowly rock backwards and forwards to reveal features at the edges, an effect known as LIBRATION.

After this time many scientists and astronomers turned their telescopes towards the Moon and made more or less successful attempts to map its surface. Of the more successful of these may be mentioned Claude Mellan, who made three engravings of various phases of the Moon around 1636 (*Plate* 2) which are undoubtedly the most lifelike representations of the seventeenth century. Complete maps were published by a number of people, but those of Cassini in 1680 (*Plate* 3A) remained the most detailed and accurate for 100 years. A globe of the Moon, showing the features of its surface in relief, was made by Sir Christopher Wren for Charles II in 1661, at the request of the Royal Society, though it is unfortunately now lost.

The first names were assigned to lunar features by Langren, who named them after towns, philosophers and popes of the time. Only three of the names he proposed survive today: Sinus Medii, Langrenus (after himself) and Pythagoras. The present system of naming craters after notable astronomers, scientists, writers, etc., particularly those connected with lunar study, was devised by Riccioli in 1651.

Some of the names used by the early lunar cartographers, such as 'Vesuvius' and 'Etna', suggest that they assumed the lunar craters were volcanic, but the first definite attempts at explaining the origin of lunar craters were made by Robert Hooke in his *Micrographia* of 1665. In some experiments with boiling alabaster he observed that bubbles of air

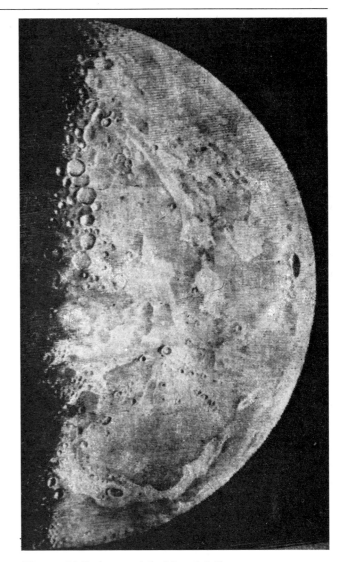

Plate 2 Mellan's map of the Moon (1636).

rose and burst, leaving craters very like those of the Moon. He suggests that these might be analogous to both lunar craters and volcanoes on Earth. He also suggests that the craters might have been formed by the impact of bodies falling on to the Moon, but since asteroids were not discovered until the early nineteenth century, and meteorites were not accepted as extraterrestrial until about the same time, he rejected this hypothesis outright.

Until the nineteenth century it was generally accepted by all lunar authorities that lunar craters were volcanic, as volcanology was not in such an advanced state that the vast differences between many lunar craters and volcanic craters could be detected. The large dark areas, now known to be lowlands, were thought to be seas by many of the

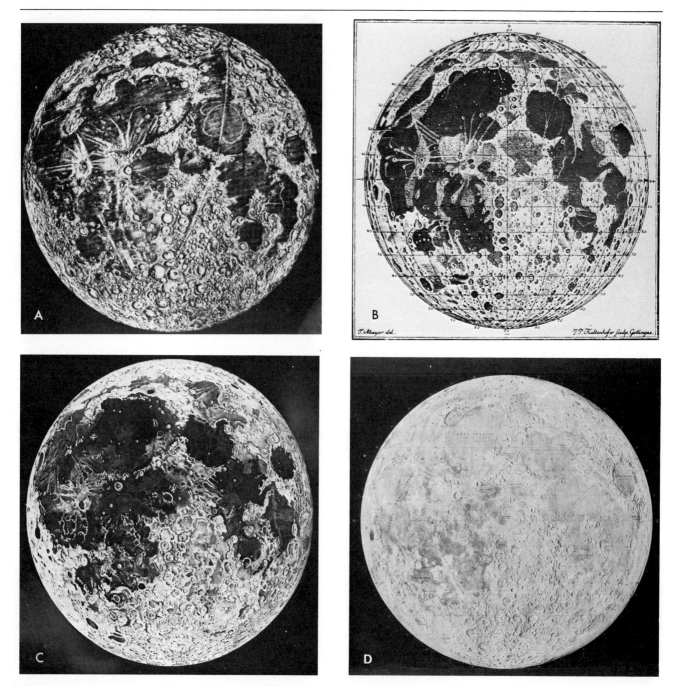

Plate 3 Comparison of four lunar maps by: (A) Cassini (1680); (B) Tobias Mayer (from observations 1750); (C) Russell (1805–6); and (D) American ACIC chart (1962).

early lunar observers, an idea also expressed by Leonardo da Vinci in the sixteenth century from his observations made with the naked eye.

Lunar study after Cassini received no significant advances until Tobias Mayer (1723–62) made the first accurate positional measurements of lunar craters around 1750. He died before he could complete a map from his observations, however, but a map was prepared from his work and published in 1775 (*Plate* 3B).

The most lifelike drawings and maps of the Moon before the invention of the photograph were undoubtedly those of an English portrait artist, John Russell, RA (1745–1806). For 40 years he had a spare-time hobby of making sketches of the Moon whenever possible, and supplemented these with

Historical Introduction

a large number of micrometrical measures with which he built up a triangulation grid of points on the lunar surface. He used these to make his large pastel drawing of the Moon in 1795 (*Plate* I), and also his globe of 1797 and his map and full Moon drawing of 1805 (*Plate* 3C). He had the advantage over other lunar cartographers in that he was a fully accomplished professional artist, and his map, drawn so that each portion of the Moon appears under a similar oblique illumination, stands excellent comparison with the recent NASA lunar photomosaic of 1962 (*Plate* 3D). His ingenious lunar globe also contained a mechanism, designed by himself, for demonstrating the libratory motions of the Moon with some accuracy.

Possible changes on the Moon were frequently remarked on about this time, and Sir William Herschel (1738–1822) created something of a stir by observing what he thought were volcanic eruptions on the unlit portion of the Moon. In April 1787 he wrote: 'I perceive three volcanoes in different places on the dark part of the new Moon. Two of them are already nearly extinct, or otherwise in a state of going to break out; the third shows an eruption of fire or luminous matter.' Schröter (1745–1816) was of the opinion, probably correct, that Herschel was observing bright craters illuminated by EARTHSHINE.

The possibility of life on the Moon was universally entertained by scientists of this period, and Herschel at one time stated that life on the Moon was 'an absolute certainty'. It was only towards the end of the nineteenth century as more physical observations of the surface were made that the possibility was generally dismissed.

Throughout the nineteenth century improvements in lunar cartography continued as more and more detailed surveys appeared. One authority, Schmidt, was also the cause of a remarkable stir in 1866 when he announced that the crater Linné, which he had been familiar with since 1841 as a deep crater, had disappeared and was replaced by a whitish cloud. This alarming announcement caused much searching of the literature, and it was found that Schröter's observations round about 1788 corresponded with the 1866 appearance of the feature. On the other hand, Lohrmann and Madler refer to Linné as a deep crater, in terms inconsistent both with Schröter's observations and the present-day aspect. Linné is a tiny formation, however, and there is little doubt that the discrepancy between the observations is due to their being at the limits of the instruments used, though at the time the observations were treated seriously as a vast unexplained mystery.

The development of photography in 1837 and its first application to the Moon in 1840 allowed a significant advance in lunar cartography, though photographs did not achieve the resolution of existing maps until the end of the century, when the first photographic atlases were compiled.

Visual observations continued to be made alongside the improving photographic results, though the two types of observation, with better resolution on the one hand and more accuracy on the other, were not combined to full advantage until the American ACIC charts of the lunar surface, which combined visual and photographic observations at a scale of 1:1,000,000, were published between 1962 and 1967.

The stimulus of the Russian and American space programmes on the study of the Earth and the Moon, and the vast increase in knowledge of both bodies as a result of these programmes, is well known and is described in some detail in the chapters that follow.

Origin of Lunar Features

Alongside these advances in knowledge of the lunar surface through the centuries, theories of the origin of the Moon and her features have developed also, particularly during the past 100 years. Most famous and controversial is the question of the origin of the lunar craters. Since Gruithuisen proposed the impact hypothesis in 1829 controversy has raged between its supporters and those of the volcanic hypothesis, though the close-up pictures from Orbiter probes suggest that a high proportion of lunar craters are of impact origin. Most authorities now agree that both types of crater occur on the Moon.

Other theories of a greater or lesser degree of crankiness have been advanced from time to time: El Campo has suggested that the craters are the result of bomb explosions in a war between two races of lunar inhabitants, causing the lunar seas to be blown into space and eventually to fall back on the Earth, causing the biblical Flood. Weisberger regarded craters as storms and cyclones in the lunar atmosphere, and Beard thought they were coral atolls formed in an immense lunar ocean

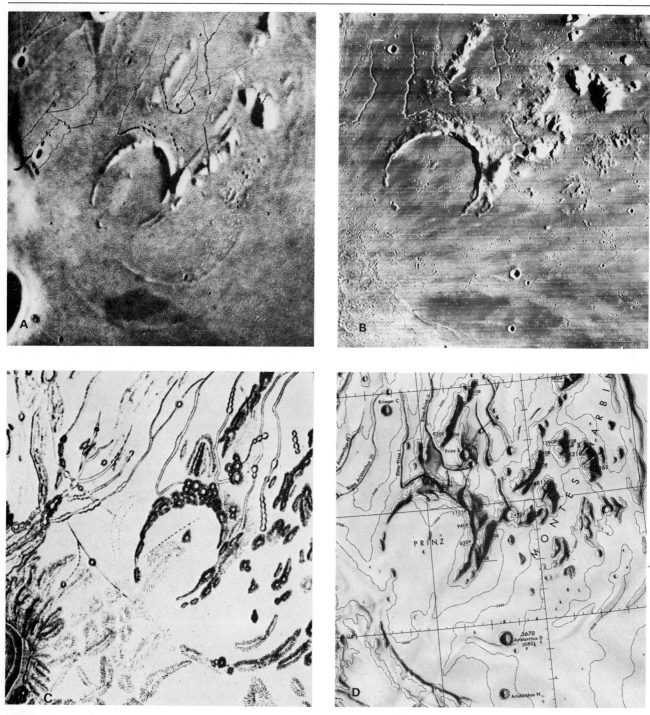

Plate 4 Four views of the lunar crater Prinz: (A) from Krieger's *Mond Atlas* (1912); (B) an Orbiter picture; (C) map by Fauth (1932); and (D) American ACIC chart.

before it dried up. Boneff suggested that the gravity of the Earth sucked molten matter in the interior of the Moon through weak spots in the crust, the resulting holes forming the craters.

Finally, not a superstition or erroneous idea but an extraordinarily successful hoax concerning the Moon occurred in 1835, the central figure being Sir John Herschel, son of Sir William. R.A. Locke, journalist with the *New York Sun*, took advantage of Sir John's absence in South Africa to invent a remarkable story about some supposed astronomical discoveries made there by him with a new gigantic telescope of a revolutionary new design, using electric light to supplement the feeble light received from the Moon—a principle which, of course, is optical nonsense. With this it was said he was able to bring the Moon to an apparent distance of 100 yards. In the *Sun* of Tuesday, 28 August 1835 Locke tells of the supposed observations when this imaginary instrument was turned towards the Moon. He describes in great detail the hills, vales, vegetation and forests that were observed, and then living creatures: bison-like and bear-like creatures and finally intelligent creatures . . . 'like human beings . . . erect and dignified. They averaged four feet in height, were covered, except on the face, with short and glossy copper-coloured hair, and had wings composed of thin membrane, without hair, lying snugly upon their backs, from the tops of their shoulders to the calves of their legs.' Needless to say, the *Sun* rapidly hit the top of the world newspaper circulation, and on the day of the 'batman' story sold every newspaper that the presses could turn out in 10 hours of running.

Incredibly, America was completely taken in. One man claimed to have seen the great lens of Sir John Herschel's telescope before shipment in England, and another that he knew the story was true because he had a copy of the *Edinburgh Journal of Science*, a fictitious journal invented by Locke and from which the account was supposed to have been taken. One ladies' club even enquired whether it might be possible to communicate with the lunar beings in order to convert them to Christianity. Europe, however, remained incredulous, and shortly afterwards Locke admitted to a friend that he had been the author of the whole thing, and the friend, being on the staff of a rival New York newspaper, immediately discredited the story. Nevertheless Locke maintained some esteem for years afterwards as the man who took a whole continent for a ride.

The Study of the Earth Since the Renaissance

Knowledge of the Earth has obviously grown in a different manner from knowledge of the Moon. With the Moon, men began with a general view of the whole globe and strove to obtain a closer and closer view of smaller areas. Knowledge of the Earth, on the other hand, began with detailed knowledge of small areas and has gradually extended to a more and more general view of the whole globe.

Leonardo da Vinci, in the fifteenth century, made a number of drawings and notes of the rocks he observed and the fossils he found in them, and attempted some kind of rudimentary classification, but theology provided a great hindrance to the study of the Earth until the second half of the eighteenth century. The events of Genesis were generally accepted as true, and Archbishop Usher (seventeenth century), by studious deductions based on the events recounted in the Old Testament, calculated the world creation to have occurred in 4004 BC. Detailed knowledge of some rock types was growing in the eighteenth century under the stimulus of mining interests, but any deviation from biblical teaching met with strong opposition. Nevertheless, the vast apparent upheavals hinted at by many geological structures attracted the notice of some freethinkers. Early ideas included those of floods and catastrophes to explain the features they observed. Werner (1749–1817), a German at Freiburg, thought that a series of gigantic major floods had *precipitated* rocks in the observed stratified layers, an idea that came to be known as Neptunism. Cuvier, a Frenchman and a great palaeontologist, talked of successive creations after each flood.

The first real advances were made by James Hutton (1726–97), a farmer from Edinburgh, who was the first to look for generalisations from field evidence, and thus the first to apply the scientific method to geology. Under pressure from his friends he published his theories in a short paper to the newly formed Royal Society of Edinburgh in 1785. This paper was ignored, then violently attacked five years later. Eight years later Kirwan, an Irishman, published an even more severe attack, misrepresenting Hutton's views and calling him an atheist. This inspired Hutton to write a far more complete work, published under the title of *Theory of the Earth, with Proofs and Illustrations*. John Playfair, a great friend and supporter of Hutton's views, expanded and popularised them in a far more readable volume: *Illustrations of the Huttonian Theory of the Earth*.

The central idea of Hutton's whole theory is that 'the present is key to the past'. He explains all observable features on the Earth not by huge catastrophes and different conditions but by processes observed to occur now. This principle came to be known as *Uniformitarianism*. He saw these processes acting steadily through time with, to use his own poetic phrase, 'no vestige of a beginning, no prospect of an end'. He was the first to discover a number of important geological and geophysical principles. He realised that sedimentary rocks were for the most part compacted sea-bed deposits, he thought that the consolidation of rocks was due to subterranean heat and that the folds of the strata were due to deep-seated earth movements. He also recognised UNCONFORMITIES as being the dividing line between two different periods of deposition, with a large space of time between. He was the first to recognise that volcanoes and lava flows were not burning in the normal sense of oxidising, and suggested that the nucleus of the Earth might be 'a fluid mass, melted, but unchanged by the action of heat'. He distinguished INTRUSIVE ROCKS, recognising DYKES for what they were, and that granite was not the 'primeval chemical precipitate' that prevalent opinion of the time dictated but was intruded into rocks after they had been laid down. He also noticed METAMORPHISM of the rocks adjacent to igneous intrusions. On the erosion of landscapes, he realised that erosive processes were always at work, and that stream valleys were eroded by the streams themselves, a theory hotly disputed until 80 years after Hutton's death.

The great success of all Hutton's theories lay in the fact that he stuck to his principle of invoking only known processes, and did not wander in unfounded fantasies.

William Smith, in the early nineteenth century, drew together into one single geological map of England the detailed knowledge of rocks gained in several different smaller areas. He stressed throughout his life the importance of geological maps as a vital tool in geology, and was the first to observe that each geological age has its own characteristic assemblage of fossils and to use these for dating rocks. Much of the energies of geologists since his time have been devoted to extending and improving geological maps to cover the rest of the Earth, much of which remains unmapped even now. By the end of the century four major ideas had been accepted: Uniformitarianism; the unlimited time of geological processes (especially after Darwin); the importance of observation, not revelation; and that landforms change through time.

Like all other fields of knowledge, the study of the Earth has accelerated rapidly during this century. In the past five years the space vehicles that have given us much closer views of the Moon have given us views of the Earth from much farther away, revealing not only weather patterns more clearly but also fundamental geological structures not guessed at before. In the past, any kind of comparison between the Earth and the Moon has been hindered by the distance of the Moon and the proximity of the Earth. Only now that we have reached great distances from the Earth, and actually stood upon the Moon and brought back pieces of it, can we begin to make some kind of real and fruitful comparison between the two worlds.

Plate I
Russell's pastel drawing of the Moon (1795).

Plate II
Aconcagua, the highest mountain in the Andes of South America, rising to some 7,000 metres above sea-level. The pyramidal form of this mountain is typical of a glacially eroded mountain. *Photograph: J. E. Guest.*

Plate IIIA
Glaciers in the Huascaran mountains of Peru. The blue lake in the deep glacier-cut valley is filled with meltwater from the glaciers. It probably owes its blue colour to a suspension of fine rock powder produced in the glaciers by grinding together of the rocks. *Photograph: J. E. Guest.*

Plate IIIB
A glacier on the Columbia Ice Field, Jasper National Park, Canada. Note the crevassed nature of the glacier's surface.
Photograph: R. Mason.

Plate IIIC
Folded rocks forming mountains of the Glärner Alps, Switzerland.
Photograph A. J. Lloyd.

Chapter Two

Astronomy of the Earth-Moon System

E. L. G. BOWELL

The Moon is the only large natural satellite of the Earth. In contrast with the satellites of other planets it stands out immediately by virtue of its great size and mass, ranking sixth largest of the 32 planetary satellites. It is easily the largest with respect to its primary planet (*Table* 1).

Both the Earth and the Moon are very nearly spherical. It would be most remarkable if they were not since, according to theory, any very large surface structures ought to collapse or sink under their weight. However, there are two dynamical effects that distort the globes of the Earth and Moon: ro-

TABLE 1

Primary planet	Mass (tons)	Number of satellites	Mass of largest satellite (tons)	Ratio of satellite's to planet's mass	Ratio of satellite's to planet's diameter
Earth	6.0×10^{21}	1	7.3×10^{19}	1:81	1:4
Mars	6.5×10^{20}	2	1.3×10^{13}	1:50,000,000	1:300
Jupiter	1.9×10^{24}	12	1.5×10^{20}	1:13,000	1:28
Saturn	5.7×10^{23}	10	1.4×10^{20}	1:4,100	1:25
Uranus	8.7×10^{22}	5	2.7×10^{18}	1:33,000	1:47
Neptune	1.0×10^{23}	2	1.4×10^{20}	1:710	1:11

The Earth is the only planet with a single satellite, and this idiosyncrasy, coupled with the relative sizes of the Earth and the Moon, has led some scientists to call the Earth-Moon a twin planet. However, this term is not necessarily a good one since it implies a genetic connection. Table 2 gives some comparative properties of the Earth and Moon.

tation and tides. A rotating planetary body bulges at its equator. In the case of the Earth the equatorial diameter exceeds the polar diameter by 43 km. The Moon's smaller size and longer rotation period of 27.3 days endow it with an equatorial bulge estimated to be only 16 metres—far too small to be detectable.

The effects of tidal distortion on the Earth and

TABLE 2

	Equatorial diameter (km)	Surface area (Earth = 1)	Volume (Earth = 1)	Density (g/cm³)	Rotation period (days)	Surface gravity (Earth = 1)	Escape velocity (km/sec)
Earth	12,756	1.000	1.000	5.52	0.99	1.000	11.2
Moon	3,476	0.075	0.020	3.34	27.32	0.165	2.4

The Earth and Its Satellite

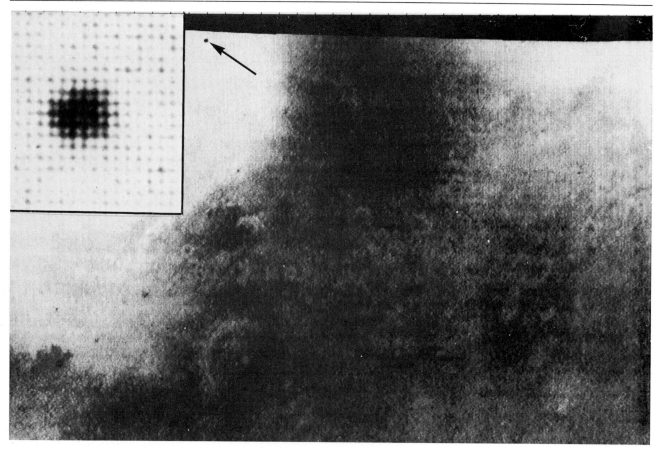

Plate 5 Mars Mariner picture showing the satellite Phobos silhouetted against the Martian surface. The insert (*top left*) is an enlargement from the same picture to show the non-spherical form of this satellite. *NASA photograph.*

the Moon are likewise small. We observe the tide-raising force of the Moon as a heaping up of the oceans by several metres on the sides of the Earth directed towards and away from the Moon. This is a more complicated situation than would arise if the whole bulk of the Earth, its oceans and atmosphere were distorted equally, in which case ocean tides would not be directly apparent to us. In fact, the solid bulk of the Earth does suffer a small tidal distortion, amounting to a rise and fall of about 1 metre.

The Earth has a tide-raising effect on the Moon which is 81 times larger than the Moon's tidal pull on the Earth (in proportion to the Earth-Moon mass ratio). Since the Moon always presents the same face to the Earth the resulting tidal bulge should be 'frozen' in the bulk of the Moon. According to theory it should give rise to a 95-metre difference in the Moon's equatorial and polar diameter —again too small to be measured. The Sun also exerts a tide-raising influence on both the Earth and Moon equally; its effect is a little over half that of the Moon on the Earth.

To generalise for the whole solar system we can say that the overall shapes of the planets, their satellites, the ASTEROIDS and smaller débris in orbit around the Sun are controlled by three principal influences: mass, size and rate of rotation, and tidal effects of neighbouring bodies. The relatively massive bodies (planets and larger satellites, including the Moon) do not behave like rigid globes, and large surface height differences (mountains and troughs) cannot be maintained for long periods. Smaller satellites and asteroids may, however, depart radically from spheres. Direct evidence that Phobos, one of Mars' tiny satellites, is not spherical has recently been obtained by the Mariner VI and VII Mars fly-by missions. Photographs show the outline of Phobos to be irregular, measuring 22 by 18 km. (*Plate* 5).

The Earth-Moon system is so different from all

the other planet-satellite systems in the solar system that it must be thought of as an abnormal system. The 'normal' satellite systems are those of Jupiter, Saturn and Uranus, where all the satellites are very small compared with their primary planets and move in almost circular orbits above their planets' equators.

The Orbital Paths of the Earth and Moon

To a rough approximation the Earth pursues an elliptical path around the Sun, and the Moon attends it, making about 13 revolutions annually in orbit around the Earth. Happily for life on our planet, the Earth's distance from the Sun varies by only 2,500,000 km about its mean value of 149,599,000 km. The Moon's orbit about the Earth (*Figure* 1) is likewise approximately elliptical, but a little less circular than the Earth's about the Sun.

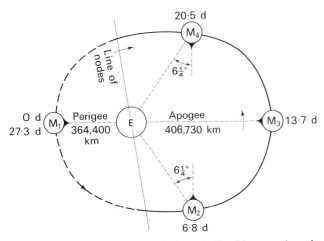

Figure 1 *The orbit of the Moon relative to the Earth* is approximately an ellipse with the barycentre at one focus. The Moon is shown at quarter-month intervals (M_1, M_2, M_3, M_4), and the phenomenon of libration in longitude is illustrated by the $6\frac{1}{4}°$ disk displacement at M_2 and M_4.

Its PERIGEE distance (minimum distance from Earth) is 366,400 km and its APOGEE (maximum) distance is 406,730 km; hence there is a total distance variation of about 10% during the lunar month. This is reflected in a corresponding change in the apparent size of the Moon seen from Earth.

The Moon's Orbit

The path of the Moon in space around the Sun is not, as might at first be imagined, a spiral one focused on the Earth. Instead the Moon moves in a trajectory which is everywhere concave to the Sun (*Figure* 2). As far as the lunar environment is concerned we can take it that the Moon's solar orbit during the course of a year is very similar to that of the Earth.

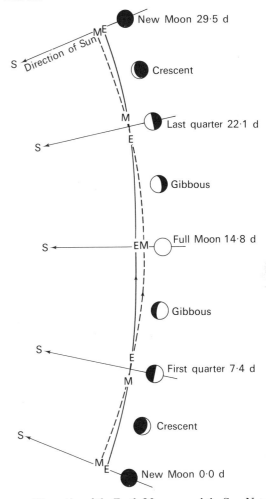

Figure 2 *The motion of the Earth-Moon around the Sun.* Note that the Moon's path (dotted) is everywhere concave to the Sun, the Earth-Moon barycentre pursues an elliptical orbit. The phases of the Moon as seen from Earth are shown with their average intervals (the Moon's age) after new Moon.

However, our picture of the Moon in an elliptical orbit around the Earth, and of the Earth-Moon in an elliptical orbit around the Sun, is an oversimplification for several reasons. Rather than the Moon moving in orbit about the centre of the Earth we ought to say that the Earth and the Moon are orbiting each other. During the course of a lunar month both the Earth and the Moon move around their BARYCENTRE: the centre of mass of the Earth-Moon system. If we imagine holding a model of the Earth-Moon system, two spheres connected by a rigid rod like a very long dumb-bell (*Figure* 3), then the barycentre would be the point of balance. And since the Earth is 81 times more massive than the Moon we should find this balance point to be situated quite close to the centre of the Earth—about 1,650 km below its surface. The Earth performs a small elliptical orbit about the barycentre which is a miniature replica of the Moon's orbit.

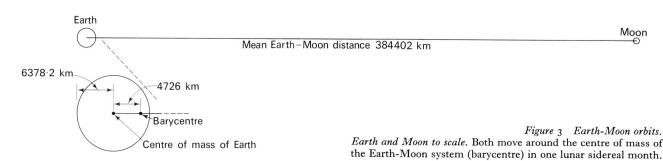

Figure 3 Earth–Moon orbits. *Earth and Moon to scale.* Both move around the centre of mass of the Earth-Moon system (barycentre) in one lunar sidereal month.

The Inclination of the Moon's Orbit

The Moon's orbit about the Earth is inclined by about 5° to the Earth's orbit around the Sun. Since this angle of inclination is quite small, the Moon's apparent monthly path among the stars is fairly similar to the Sun's annual path (the ECLIPTIC). Both bodies move through the ZODIAC constellations. If the two orbits coincided, so that the Moon and Sun traversed exactly similar paths in the sky, then there would be two eclipses every month: an eclipse of the Sun when the Moon passed between the Earth and the Sun, and an eclipse of the Moon when the Earth passed between the Moon and the Sun. However the 5° inclination causes the Moon's orbit to intersect the ecliptic, and the Moon crosses the ecliptic twice per month at the LINE OF NODES (*Figure* 1). This imaginary line does not lie in a fixed direction in space, but moves slowly westwards through the zodiac, making one complete circuit through all 12 constellations in 18 years.

Phases of the Moon and Earthshine

The phases displayed by the Moon depend on the relative positions of the Sun, Moon and Earth. Figure 2 shows the changes in aspect of the illuminated disk during the course of a SYNODIC lunar month. At new Moon, when the Moon's *age* is said to be 0 days, the Moon appears in the same direction in the sky as the Sun, so that the face presented to the Earth is unlit. During the course of the month the Moon moves eastwards in the sky with respect to the Sun and stars. Soon after new Moon a narrow crescent becomes visible, increasing in width until, at 7·4 days old, when the Moon is 90° east of the Sun, a half-illuminated disk can be seen: first quarter. The illuminated fraction continues to increase (*wax*), and at an age of 14·8 days the whole disk is visible: full Moon. At this time the Sun and Moon are opposite each other in the sky (thus the Moon rises when the Sun sets and vice versa). Thereafter the illuminated fraction decreases (*wanes*). When the Moon is 22·1 days old its disk is once more half illuminated: last quarter. And after 29·5 days the Moon is once more new. Between last quarter and first quarter the phase is said to be *crescent*; between first quarter and last quarter the phase is *gibbous*.

A moment's reflection will reveal that an observer on the Moon's earthside surface would see the Earth displaying complementary phases during the course of a lunar month. That is, he would see 'full Earth' when we see 'new Moon', and so on (*Plate* 67). It follows that near to the time of new Moon, when we see a thin sliver of a crescent, a lunar observer would see a bright, almost fully illuminated Earth. This terrestrial illumination of the unlit part of the crescent Moon is known as *earthshine*, and it was studied, before the epoch of extraterrestrial flight, to estimate the Earth's appearance from space.

Eclipses of the Sun and Moon

Eclipses, either of the Sun by the Moon (*Plate* 6) or of the Moon by the Earth, occur when the Sun, Earth and Moon lie in a straight line. This situation can arise only at the time of new or full Moon when the Moon is fairly near the line of nodes in its orbit. Figure 4 illustrates the geometry. In Figure 4*b* the Moon is exactly interposed between the Earth and

Astronomy of the Earth-Moon System

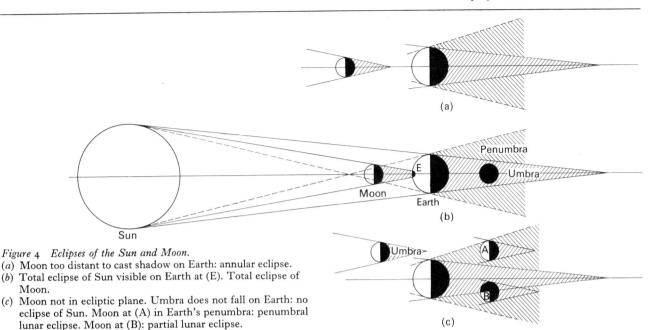

Figure 4 *Eclipses of the Sun and Moon.*
(a) Moon too distant to cast shadow on Earth: annular eclipse.
(b) Total eclipse of Sun visible on Earth at (E). Total eclipse of Moon.
(c) Moon not in ecliptic plane. Umbra does not fall on Earth: no eclipse of Sun. Moon at (A) in Earth's penumbra: penumbral lunar eclipse. Moon at (B): partial lunar eclipse.

Sun, and casts its shadow (*umbra*) on a small circular region on Earth (E). An observer at E would see the Sun totally eclipsed by the Moon (*Plate* 6). From a location adjacent to E only a partial solar eclipse would be seen, the Moon covering a part of the Sun's disk. Also shown in Figure 4b is the Moon totally eclipsed by the shadow of the Earth.

However, even when the Sun, Moon and Earth lie exactly in a straight line there need not be a total eclipse of the Sun. If the Moon is near its maximum distance from the Earth it can be too remote to cast a shadow on the Earth's surface. In this case the eclipse is *annular*, and the bright circular edge of the Sun rings the black disk of the Moon (*Fig.* 4a).

Figure 4c illustrates what happens when the Moon is not quite in the plane of the ecliptic. First there need not be any solar eclipse at all, in which case the Moon's umbra misses the Earth entirely. Likewise a lunar eclipse need not occur, or else the eclipse might be penumbral (A) or partial (B). An observer on the Moon at the time of a penumbral lunar eclipse would see the Earth partially covering the Sun's disk (he would call this a partial solar eclipse). Clearly there may be any number of gradations in the type of eclipse that can occur; the precise circumstances of each one depend on the orbital configurations of the Earth, Moon and Sun.

The duration of any eclipse is similarly governed by the motions and positions of the Earth, Moon and Sun. In a total solar eclipse, for example, the circular shadow of the Moon (projected on the Earth's surface at E in Figure 4b) tracks across the Earth's disk at about the orbital speed of the Moon (1 km/sec). The ground speed may vary from 470 metres/sec upwards owing to the Earth's sphericity and rotation about its axis, so that an observer in the *path of totality* will experience a total solar eclipse for a short time only. The *duration of totality* cannot exceed 7 minutes 30 seconds, in which case the path of totality would have a maximum width of 260 km. Lunar eclipses may be total for up to 100 minutes since the diameter of the Earth's umbral shadow can be as great as 10,000 km at the distance of the eclipsed Moon.

There are limits on the number of eclipses in any year: at least two but not more than seven can occur. On average there are about 230 total solar and 150 total lunar eclipses per century.

During every eclipse many astronomical experiments are carried out. Important from our present point of view are temperature measurements made on the eclipsed Moon (Chapter 8).

The Rotation of the Moon about its Axis

It is a common observation that the Moon always presents the same face to the Earth, and therefore does not rotate about its axis with respect to the Earth. Yet it does rotate with respect to the stars and the Sun. The periods of orbital revolution and axial rotation are equal, and the Moon is said to rotate synchronously (*Figure* 1).

An observer on the earthside hemisphere of the

Plate 6 An eclipse of the Sun.

Moon would note that during a lunar SIDEREAL month (27·3 days) the stars make a complete circuit about the sky, the Sun a little less than one circuit (since it moves eastwards among the stars, reflecting the motion of the Earth-Moon in solar orbit), whereas the Earth would remain almost fixed in direction with respect to the horizon. On Earth we are used to the Sun and stars making a complete circuit of the heavens in 24 hours, rotating about a fixed point marked approximately (in the northern hemisphere) by the Pole Star. On the Moon the stellar motion around the heavens would appear to proceed 27·3 times more slowly, and the fixed pole of rotation would not be situated near the Pole Star but about 25° away in the constellation Draco.

Figure 5 shows the relative orientations of the principal planes and axes in the Earth-Moon system. The plane of the Moon's equator is inclined at an angle of 6° 41′ to the plane of its orbit around the Earth, which is in turn inclined at 5° 9′ to the ecliptic. The difference between these two angles, 1° 32′, is the inclination of the Moon's polar axis with respect to the pole of the ecliptic. The pole of the Moon's orbit and the Moon's polar axis PRECESS about the pole of the ecliptic in a period of 18·6 years. It is interesting to note that the Earth's polar axis, inclined at 23° 27′ to the pole of the ecliptic, precesses in a similar way, but so slowly that each revolution takes 25,780 years (in effect our present Pole Star, Polaris, will be more than 45° from the direction of the Earth's polar axis in 13,000 years' time, and most of the constellations making up the *zodiac* will be different from the present ones).

Most of the variations in the facts and figures pertaining to the Moon's orbit are due to the equatorial bulge of the Earth and the inclination of the Moon's orbit to the Earth's equator. Additionally the Sun plays a major role in perturbing the whole Earth-Moon system.

Libration

The Moon rotates about its axis at a uniform rate, but does not orbit the Earth at a uniform speed.

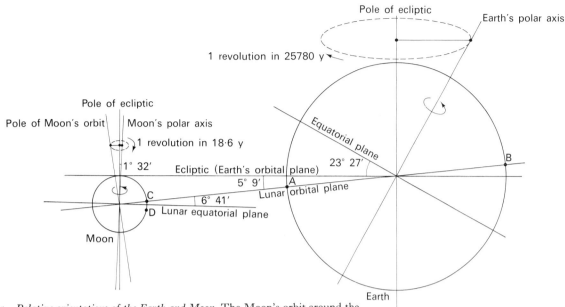

Figure 5 Relative orientations of the Earth and Moon. The Moon's orbit around the Earth is inclined at 5° 9′ to the Earth's orbit around the Sun. On Earth the Moon is overhead at (A), and the centre of the Moon's disk (C) does not coincide with the lunar equator. This is the effect of lunar libration in latitude (shown here at its maximum value). The pole of the Moon's orbit, together with the Moon's polar axis, precess about the pole of the ecliptic in a period of 18·6 years.

This, and the inclination of the Moon's orbit to the Earth's equator, give rise to the effects known as *optical libration*.

Figure 1 shows the Moon's position in orbit around the Earth at quarter-month intervals. Starting at perigee with the Moon at M_1 we can investigate what happens to a point at the centre of the Moon's disk (shown as a peak on the Moon, facing the Earth at perigee). After a quarter of a sidereal month has elapsed (6·8 days) the Moon is at M_2, having turned 90° with respect to the stars. An observer on the Earth would notice the peak displaced about $6\frac{1}{4}°$ to the left (west) on the Moon's face because the Moon moves more than a quarter of the way along its orbit in this time. At apogee (M_3) the centre of the lunar disk and the peak once more coincide. At M_4 (20·5 days, three-quarters of the month elapsed), the peak would appear to be displaced by $6\frac{1}{4}°$ to the right (east). At M_1 (27·3 days) the cycle recommences. The effect seen from Earth (called the optical libration in longitude) is therefore a slight east–west oscillation of all features on the lunar disk about their mean positions.

There exists also an optical libration in latitude: the lunar disk oscillates about its mean position in a north–south direction. In Figure 5, an observer on the Earth at A will see the centre of the Moon's disk occupied by C. In the orientation of the Earth-Moon system shown, the point C has a lunar latitude of 6° 41′ N. The Moon is at its maximum distance below the ecliptic plane. When it has moved through half of its orbit it will occupy a position above the ecliptic plane (off the diagram to the right of the Earth). At that time an observer on the Earth at B would see the centre of the disk occupied by D, a point on the Moon with lunar latitude 3° 37′ S.

The oscillations in lunar longitude and latitude are independent, but both take place with a period of about one lunar sidereal month. Other phenomena give rise to displacements of the Moon's disk but these will not be described here.

Two important consequences ensue from the libration effects. First, they enable us to see over half the lunar surface from the Earth—about 59% of its total area. Secondly, they help us to determine the shape of the Moon.

Finally, to illustrate the effects of libration, it is instructive to work out the apparent motions of the Earth and Sun when seen from the lunar surface. Figure 6 illustrates these from a vantage-point at the Moon's north pole.

During the month the Earth appears to move along a curved path—a direct consequence of libration—displaying a complete cycle of phases. Its disk is above the horizon for about 14 days and takes almost a day to rise or set. In the month chosen for the diagram (April 1968) the Sun is only partly above the horizon. From the Moon's north pole it would appear to skirt the horizon, making one circuit in 29·5 days, and the stars would turn about a point directly above the observer's head once in 27·3 days. Soon after the Earth rises, an eclipse of the Sun takes place. At 13 days 2 hours the Sun goes into eclipse behind the 'new Earth', emerging at about 13 days 7 hours. The Earth steadily rises during this interval, and for most of the rest of the month it moves in a curve above the horizon, changing phase from 'new Earth' to 'full Earth' (27 April) in the interval. Also shown in the diagram is the changing orientation of the Earth's polar axis due to the inclination of the Moon's orbit to the Earth's equator. In this particular month only the north pole of the Earth is sunlit, but it can be seen tilting towards (5 April) and away from (17 April) the lunar observer. PERIGEE (smallest distance, largest disk) takes place on 14 April, and APOGEE (greatest distance, smallest disk) on 1 April.

The Shape of the Moon

Measuring the shape of the Moon is vital to studies of the lunar interior, and practically useful for astronavigation. We know, by observing the shape of the LIMB that differences of up to 8 or 9 km in the heights of peaks and troughs exist. It is also quite clear that the MARIA are generally depressed 1 or 2 km below the HIGHLANDS. However these are only local height differences. The problem of measuring any overall

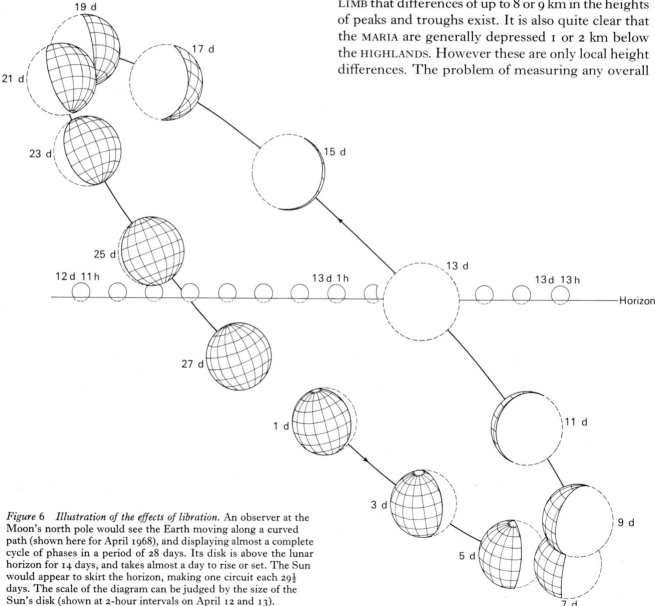

Figure 6 Illustration of the effects of libration. An observer at the Moon's north pole would see the Earth moving along a curved path (shown here for April 1968), and displaying almost a complete cycle of phases in a period of 28 days. Its disk is above the lunar horizon for 14 days, and takes almost a day to rise or set. The Sun would appear to skirt the horizon, making one circuit each 29½ days. The scale of the diagram can be judged by the size of the Sun's disk (shown at 2-hour intervals on April 12 and 13).

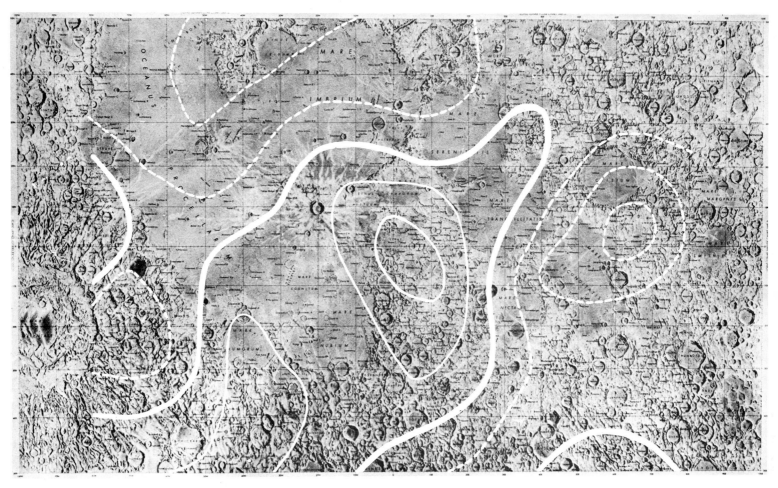

Plate 7 Map of the Moon showing global height differences. Contour interval is 1 km. The highest point is the centre of the nearside, suggesting the Moon bulges towards Earth. Thick contours indicate zero levels, dotted ones negative values.

deviation of the Moon from sphericity is extremely difficult. It is aggravated by the Moon presenting only one face to us so that the most sensitive height-mapping technique, photogrammetry, which depends upon seeing features from different angles, cannot be carried out.

Libration provides the possibility of using stereoscopic methods for objects near the centre of the lunar disk, even though the total displacement of any feature amounts to only $\pm 7°$ in latitude and longitude. The exact positions (in lunar latitude and longitude) of a number of selected features are found, then from measurement of their displacements on the disk due to libration their distances from the centre of the Moon are calculated. Up to the present this method has given discordant results, though most authors agree that global height differences amount to only 2 or 3 km. The isolevel contour map prepared by the Aeronautical Chart and Information Center, US Air Force, is shown in Plate 7. The heights on such a map as this can be expressed as an equation and when this is done it is clear that the height differences have not come about by a 'frozen' tidal deformation of the Moon. The observed global non-sphericity (of about 0.2 to 0.4%) has probably been caused by other forces.

The shape of the LIMB of the Moon can be determined fairly accurately by measuring the diameter of the Moon's disk in many directions. It has an elliptical form with the shortest and longest axes differing by 2 to 3 km. The longest axis is inclined in a southwest–northeast direction at about 35° to the lunar equator. Since this axis and the lunar equator do not coincide the interior of the Moon cannot be homogeneous, or even arranged in regular shells of different densities.

What process supports the height differences on the lunar surface is not known at present. One idea is that the Moon's oblate shape was 'frozen' in at some time in the remote past when the Moon was about 160,000 km from the Earth and was subjected to tidal forces some 17 times stronger than at present. A second is that motion in the interior, caused by temperature instability, could maintain the observed non-sphericity. A third idea suggests that the infall of material on to the Moon occurred

Plate 8 Map of the Moon showing the mascons (or mass concentrations). Contour interval is 10^{-6} Moon masses. Positive gravity anomalies (higher than average gravity) occur mainly in the circular maria (e.g. Mare Imbrium). Thick contour indicates zero levels, dotted ones negative values.

selectively, so that the denser material hit the lunar far side.

The Origin of the Moon

To ask the question 'How did the Moon form?' inevitably raises much larger issues not just concerning the birth of the Moon but the whole solar system. The best way to get the problem in perspective is to make a list of all the imaginable ways that the Moon can have formed. Some of these will seem trivial or unrealistic, but the reasonable ones fall into three categories:

1. Origin by breaking away (fission) from the Earth.
2. Origin elsewhere in the solar system and subsequent capture by the Earth.
3. Origin near to the Earth, either as an independent body or as a member of a twin planet.

We shall look briefly at some of the theories of lunar origin, bearing in mind that experiments on lunar rocks brought back by the Apollo 11 and 12 missions indicate that the Moon is at least 4,500 million years old, and comparable to that of the presumed age of the solar system (4,700 million years).

The theory of origin by breaking away from the Earth was the first to be suggested. The Earth is supposed to have become unstable in its rotation, perhaps shortly after its own birth at the time when its core was forming. Rotating in 4 to 6 hours, its globe became so oblate that it exceeded a critical shape and split apart. For the theorists a difficulty arises here, for it seems likely that the embryo (or proto-) Moon should either have fallen back into the Earth or escaped to become a planet in its own right. However, it is possible that tidal effects played a great role. The distended shape of the Earth could have changed abruptly in the space of an hour, the effect being to increase rapidly the size of the proto-Moon's orbit. Calculations show that the ruptured fragment of the Earth would have a mass of about 0.1 Earth mass (the Moon's mass is about ten times less than this). But since the two bodies would still be close together, additional violent

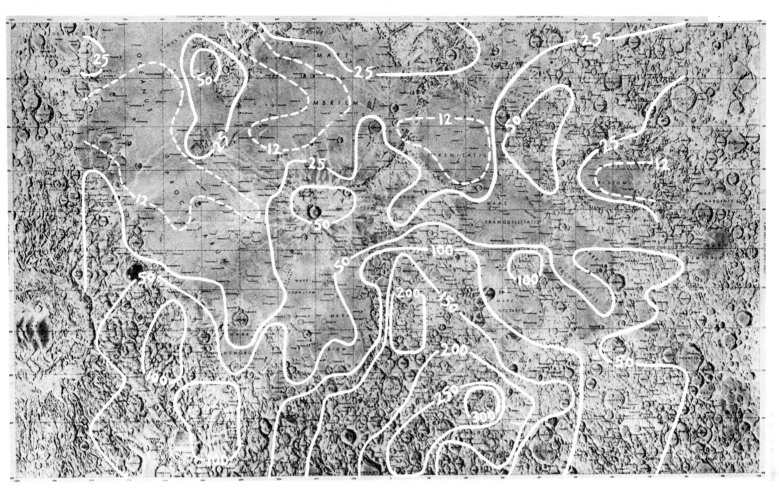

Plate 9 Map of Moon showing numbers of craters larger than 1 km across per 5,800 sq. km. The least cratering occurs in the maria suggesting them to be younger than the highlands.

events could occur. The tidal forces would be great enough to cause the surfaces of the Earth and proto-Moon to melt, and would continue to push the Moon out to a greater distance. A great deal of material from the fission process would form a bridge between the Earth and proto-Moon, and it is supposed that some of this would fall down on to the two bodies, some of it escape into space. The lunar mass could in this way be reduced to the present value. Tidal forces continued to increase the size of the lunar orbit, and to lengthen the terrestrial day.

The second type of theory proposes that the Moon was formed as a planet in its own right. It could have been a member of the ASTEROID families that orbit the Sun between Mars and Jupiter or, more likely, situated in a solar orbit quite close to the Earth's. The problem facing the theorist, that of getting the Earth to capture the Moon permanently (without resorting to the solution of having the two bodies collide), is a severe one. It is possible to calculate orbits for the Earth-Moon in which the Moon stays near the Earth for some time, perhaps a number of years, but eventually escapes. For the Moon to remain in a stable Earth orbit some external effect has to act. This might be a change in the shape of the Earth's solar orbit; or a reduction of the Moon's mean distance from the Earth, perhaps by the violent infall of meteoritic material on the Earth-Moon.

Another possibility is that the Moon approached and started to orbit the Earth in the RETROGRADE sense, spiralling in to a distance of only 18,000 km (less than 3 Earth radii). This would give rise to a very large tidal interaction, causing the inclination of the lunar orbit and the Earth's shape to change very rapidly. Eventually the Moon's orbital plane passed over the poles and the orbital motion changed from retrograde to direct. In the course of this dangerous manœuvre part of the Moon may have become detached, some of the material falling on the Earth. In this way some of the characteristic features of the Moon's nearside, such as the MARIA, might be explained.

Some theorists are suspicious that the Earth possesses just one natural satellite, noting that the

Earth is the only such planet (*Table* 1). According to them no planet can form without producing at least two offspring (or none at all). The Earth not only endeavoured, but succeeded, in the generation of three satellites during its own formation. The Moon originated at 6 Earth radii, a heavy colleague formed near the Earth's equator, and a lighter one at a distance of 34 Earth radii. The heavy satellite is thought to have reunited with the Earth, while the lighter one did not agglomerate but remained as a ring of PLANETESIMALS. The proto-Moon in its tidal retreat from the Earth collided with the planetesimals, and in this way the maria and many large craters formed more than 2,000 million years ago.

A variation on this theory, recently proposed, is that two satellites were formed by fission, and the larger one escaped into solar orbit. The Moon, in the very complicated gravitational field between the Earth and its other satellite, was retained and forced into orbit at a distance of 3 Earth radii. The escaped satellite, weighing about 0·1 Earth mass, is the planet Mars.

Astronomers are still divided as to the origin of the Moon, but the majority favour a simultaneous origin of the Earth and Moon from material in the primeval solar nebula, thereby forming close to each other in space. The analysis of trace elements in rocks brought back from the lunar surface by the crew of Apollo 11 seems to confirm this viewpoint. The DENSITY of the Moon is 3·3, and that of the Earth is 5·5. The large difference between the two has suggested in the past that the Moon was formed by fission from the Earth's crustal material. However, it is possible that during its formation the Earth swallowed up the heavy elements (thus acquiring its dense metallic core) and left the lighter materials to be incorporated into the proto-Moon. If this is so then we can rightly call the Earth-Moon a twin planet.

Chapter Three

The Interiors of the Earth and Moon

R. MASON

Introduction

Only a very small part of the Earth's interior, the highest few kilometres, is available for direct study. Men themselves have descended, in the deepest mines, some 3,000 metres below the Earth's surface, and boreholes have been put down to more than 7,000 metres. Our knowledge of the structure of the Earth at greater depths than this is based on indirect studies, from which scientists can make intelligent guesses about the composition of the Earth's interior, which in turn shed light on the way the Earth formed and has evolved since its formation. In the case of the Moon, ideas about the interior were purely theoretical up to the time of the unmanned and manned lunar landings. The information provided by the lunar landing programme is still being studied, and will give scientists a better understanding of the Moon's interior. At the moment, although ideas on the subject have developed rapidly they are still at a preliminary stage and are likely to change considerably over the next few years.

We know the masses of the Earth and Moon, and from a knowledge of their diameters we can calculate their average DENSITY. The average density of the Earth is 5·52 (the density of water is 1·0). Since the surface rocks of the Earth are much less dense than this (2·0–3·5), it follows that in the interior of the Earth there must be material of a higher density. The Moon has a lower average density than the Earth (3·33), although its surface rocks are of somewhat greater density than those on Earth, at least in the maria so far visited, so that it cannot have a very dense interior.

We can go further than this in finding out about the densities of the interiors of the Earth and the Moon from their overall physical properties. They both rotate like spinning tops about their axes, and the axes themselves have a slight wobble and do not always point in exactly the same direction. The technical name for the wobble is PRECESSION. From a study of the movement of the axes of rotation, a property called the MOMENT OF INERTIA of the Earth or Moon is calculated. This places certain limits on the way the density increases towards the centre. Any models of the interiors of the Earth and Moon, if they are to be accurate representations, must have the densities of the different parts of the interior distributed so that the moment of inertia corresponds to the correct value.

Earthquakes

One of the most important indirect indicators of the nature of the Earth's interior is the study of earthquakes, the science of seismology. Earthquakes are among the most frightening natural phenomena because they are so destructive and we have no ability to prevent or even predict them (*Plates* 15, 16). However, we are beginning to understand their causes and to construct buildings so that loss of life in even a severe earthquake may be limited.

An earthquake is exactly what its name says. The ground jolts, and witnesses of a strong earthquake have described it moving like the deck of a ship in a rough sea. This intense disturbance is limited to a small area a few kilometres or tens of kilometres across, but it may cause secondary effects as destructive as the earthquake shocks themselves: land-

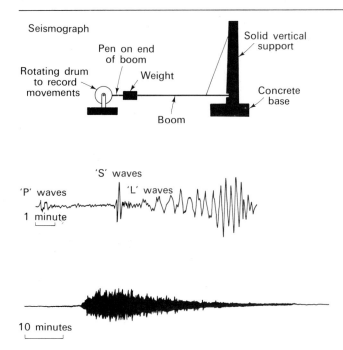

Figure 7 A Seismograph. Two seismograph traces below represent: (a) a typical large terrestrial earthquake; and (b) the quake caused by the impact of the lunar module craft into the lunar surface.

Electrical or optical instruments record the way the Earth is moving relative to the weight, and the record is usually presented as a wiggly line on a strip of paper winding round a drum at a constant speed. The amplitude (or 'height') of the wiggles of the line on the chart is a measure of the strength of the earthquake waves, and the number of wiggles occurring in each length of the chart records the number of waves arriving in each interval of time. By analysing these records, seismologists (the scientists who study earthquakes) work out the nature of the deeper parts of the Earth by methods rather like those used to work out the depth of the sea by echo-sounding.

The shock-waves produced by an earthquake are of a number of different types and each type travels at a different speed through the Earth, so that they arrive at different times at a seismograph hundreds of kilometres away from the focus of the earthquake. The first waves to arrive are called P waves (see *Figure 7*). In these waves the rock particles vibrate along the line of travel of the waves (see *Figure 8*). For this reason the P may be understood to mean 'push and pull' waves. The P waves cause a relatively small movement to be recorded by the seismograph.

Next come the S waves. In these waves the particles of rock vibrate at right angles to the direction of travel of the waves (*Figure 8*), so that the S may be understood to stand for 'shake' waves. They cause a stronger movement of the seismograph than the P waves. Last are the L waves, which travel relatively slowly and only over the surface of the Earth. They cause large movements of the seismograph and are the waves which cause the destruction in severe earthquakes. They have a longer wavelength than the P and S waves. At a large distance from the earthquake focus the L waves are delayed relative to the P and S waves, not only because they are slower but also because they have to travel to the seismograph by a longer route round the curved surface of the Earth compared with the P and S waves travelling through the interior. Seismologists recognise several types of L waves in which the particles of rock describe orbits of vibration of a circular or elliptical form.

It is the P and S waves that are important in revealing the deep interior of the Earth. One difference between the two kinds of wave is particularly significant. P waves are identical in character with sound waves in a liquid or gas and

slides, sudden floods or great waves in the sea. The earthquake is caused by sudden movements in the rocks in the outer part of the Earth (down to 700 km). These movements relieve great stresses which are present in some parts of the Earth's interior, and will be discussed further in the next chapter. The point where the movement occurs is called the *focus* of the earthquake and, as might be expected, the worst damage occurs in the area just above the focus.

From the focus shock waves travel through the Earth in all directions. Some travel deep, and it is from the study of these waves that much has been discovered about the Earth's interior. Away from the region immediately above the focus the waves quickly diminish in strength, so that they can be detected only by special instruments, called SEISMOGRAPHS. A seismograph is a very simple instrument in principle: it consists of a weight supported by a long wire or sensitive spring. As the Earth moves, the weight remains still due to its INERTIA, in the same way that a motor-car should remain steady while its wheels move quickly up and down on a bumpy road, the movement being absorbed by the car's springs and shock-absorbers. Thus, in the seismograph, the movement of the Earth relative to the weight is absorbed by the spring or wire.

The Interiors of the Earth and Moon

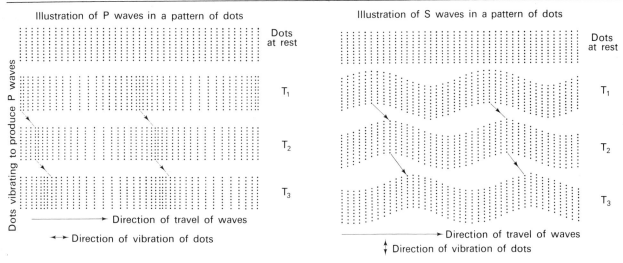

Figure 8 The form of P and S waves.

can be transmitted through any substance, but S waves can be transmitted only through solid substances having the strength to return to their former shape after being distorted by the S waves, and not through liquids.

In general, both P and S waves travel more rapidly at greater depths in the Earth. Since the speed of travel of the waves increases in denser rocks, this confirms the conclusion based on the Earth's mass and moment of inertia that the density increases with depth. However, this velocity increase is not uniform. There are marked 'discontinuities', where the velocity increases rapidly over a distance of a kilometre or so. The discontinuities act as reflecting surfaces for earthquake waves, and also deflect their direction of travel, making seismograph records more complex. Two well-marked discontinuities are particularly important, and have been named after the seismologists who first detected them. They are the

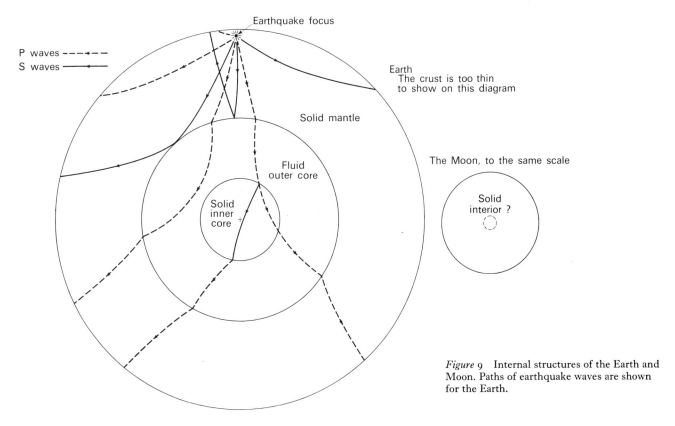

Figure 9 Internal structures of the Earth and Moon. Paths of earthquake waves are shown for the Earth.

31

MOHOROVIČIC discontinuity at a depth of 6–70 km below the surface of the Earth, and the GUTENBERG discontinuity which lies much deeper at 2,900 km. As the radius of the Earth is just over 6,000 km, this means that the Gutenberg discontinuity is not quite half-way to the centre.

The Mohorovičic discontinuity (or 'Moho' for short) divides the thin crust of the Earth, made up of rocks similar to those we see at the surface, from a deeper layer known as the *mantle*. This layer is denser than the crust. The upper part of the mantle has a density of about 3·4, compared with an average value for the crust of about 2·9. The mantle extends down to the Gutenberg discontinuity. Here a remarkable change takes place. The velocity of P waves increases sharply, indicating a sudden increase in density, from about 5·7 to 9·4. But below the Gutenberg discontinuity S waves are not transmitted. It is concluded that the layer below the Gutenberg discontinuity behaves as a liquid. This region in the very deep interior of the Earth is called the *core*. So, on the basis of earthquake-wave behaviour, the interior of the Earth is revealed as having a structure of concentric layers or shells, namely: a thin external crust and a thick mantle, which are solid; and an inner core which is, partly at least, a liquid.

Recent seismological studies have shown that there are structures within these large internal divisions. The structures of the crust and upper mantle will be discussed in the next chapter. It appears that the core has two parts: an outer zone which is liquid, and an 'inner core' which is solid. How is it possible to recognise a solid inner core if S waves cannot penetrate the outer core? When the P waves reach the inner core they produce two kinds of waves in the inner core, P and S waves. These travel through the inner core at different speeds. So at the surface two different shocks arrive which have travelled through the inner core —those which travelled as P waves in the outer core and as P waves in the inner core, and those which travelled as P waves in the outer core and as S waves in the inner core. Since they arrive at slightly different times, seismologists have been able to distinguish the two types.

The study of 'moonquakes' became possible only after seismographs had been placed on the Moon's surface by the astronauts on the Apollo 11 and 12 lunar missions. These seismographs have shown that there are fewer moonquakes than earthquakes and that they occur at or near the surface, being commonest at the point on the Moon's surface facing the Earth. To aid the seismological investigation, the discarded lunar module of the Apollo 12 mission was deliberately crashed on to the surface of the Moon to cause an artificial moonquake, and a more powerful impact was made by the final-stage rocket of the ill-fated Apollo 13 mission. Both these impacts produced remarkable traces from the seismograph, vibrating the Moon's surface for 20 minutes and an hour respectively. Comparable impacts would have produced a few seconds' vibration on Earth. The seismograph traces are quite different from their terrestrial equivalents, showing a gradual onset of vibrations, unlike the successive sudden shocks seen in terrestrial seismograph traces. This profound difference means that seismologists are unable to use terrestrial experience in interpreting lunar results, and it will be some time before any firm conclusions can be reached about the Moon's interior from these data.

Internal Composition

We can make intelligent guesses about the composition of the different layers inside the Earth, selecting substances whose physical properties, as far as we know, correspond to the properties of the deep layers. Of course, we have to allow for the very great pressures and high temperatures which must be present deep in the Earth. In this way we can arrive at a 'model' of the Earth which explains as many as possible of the facts we have about the interior.

One source of clues to guide our guesses is the fragments of matter which occasionally reach Earth from outer space, known as METEORITES. On any clear summer night, if you watch the sky for an hour or so, you will see at least one flash of light called a shooting-star, or meteor. The flash is caused by a fast-travelling particle of interplanetary matter burning by friction as it encounters the Earth's atmosphere. Most of the particles are so tiny that they are burned away before they reach the ground. However, larger particles survive the journey and reach the ground with no more than a burned crust produced by their flight through the air. The very short time taken for the flight means that the inner parts are not affected—in some cases, frost formed on recently arrived fragments which had been cold in outer space.

Plate IV
Folded layers of limestone and shale in the Jura mountains of Switzerland.
Photograph: A. J. Lloyd.

Plate V
The Sugarloaf mountain rising out of the town of Rio de Janeiro. This rock mass is an inselberg left after the surrounding rock had been stripped off by erosion.
Photograph: J. E. Guest.

Plate VI
Sief dunes in the Libyan desert. *Photograph: D. Savage.*

Plate VII
Sunset on the river Godavari, India. Such red sunsets are typical of the Indian dry season when the air has a high dust content caused by many thousands of cattle being moved back to villages in the early evening. Note the level flood plain of the river. *Photograph: A. J. Smith.*

Meteorites are of two main compositional types: stony meteorites and iron meteorites. Stony meteorites are the more common kind. They consist of SILICATE MINERALS rich in iron and magnesium. Iron meteorites are mostly composed of metallic iron and nickel. It is thought that meteorites may be fragments of a planet or asteroid which broke up, and that the iron meteorites may be fragments of the core and stony meteorites of the mantle of such a body. By analogy it is suggested that the Earth's mantle consists of silicates rich in iron and magnesium, and the core of metallic iron and nickel, the outer part being liquid, the inner part solid.

As far as the mantle is concerned, we have other clues to the composition of its upper part. Many volcanic rocks have their origin in the upper part of the mantle (see Chapter Five) and their composition suggests that they formed by the partial melting of iron and magnesium-rich silicates. Sometimes lavas from oceanic volcanoes, where the mantle is close to the surface, contain fragments of solid rock which have come from a great depth. In continental areas there are occasionally volcanic rocks originating from very great depths. These rocks often contain diamonds and are called kimberlites, after the famous diamond-mining area at Kimberley, South Africa. Kimberlites contain fragments of solid rocks which, like the diamonds, must have their origin in the upper mantle (*Plate XLII*). Finally, in deep fracture zones in the Earth's crust, rocks are sometimes brought to the surface by movements on the fracture which may have originated in the upper mantle.

All the rocks which may have come from the upper mantle have similar characteristics. They are dense enough to be mantle material, and often they contain minerals which can only be produced artificially under great pressures and at high temperatures. Compared with crustal rocks they are rich in magnesium and iron and poor in aluminium, sodium and potassium. Although the precise composition of the upper part of the mantle is uncertain (it is probably slightly different in different parts of the Earth) it seems likely that it has a similar composition to these fragments of rock, and to some stony meteorites. Since there are no marked discontinuities in the mantle, although its properties do show some changes with depth, it probably consists of stony material right down to the boundary with the core. It is highly probable that under the great pressures in the lower mantle the minerals of the upper mantle are unstable, and their place is taken by minerals in which the atoms are more densely packed together. We know little about the behaviour of rocks under these conditions, and any theories about the nature of the lower mantle must be speculative.

Magnetic Fields

High-pressure experiments suggest that the nickel-iron alloy of the iron meteorites would be liquid under conditions corresponding to those in the outer part of the Earth's core. We also have evidence that the core is a good conductor of electricity. This evidence is provided by the study of the Earth's magnetic field.

The Earth is unique among the nearer bodies of the solar system in possessing a relatively strong magnetic field. The Moon, Mars and more surprisingly Venus appear to have only weak magnetic fields, although there is evidence from the Apollo rocks that the Moon's field may have been stronger in the distant past. The Earth's magnetic field is of practical benefit in permitting navigation by magnetic compass and also, by trapping charged particles coming from the Sun (the 'solar wind'), protects us from damaging radiation.

The temperature of the mantle and core is far too high for the Earth's magnetic field to be due to the kind of permanent magnetism shown by iron at the surface of the Earth. The fluid metal of the core is stirred under the influence of temperature differences and the Earth's rotation. In the presence of a magnetic field a conducting fluid in motion, such as the core, can act as a dynamo, generating strong electric currents. Since these electric currents can in turn generate a magnetic field, as in an electro-magnet or SOLENOID, the core may be described as a self-exciting DYNAMO. Self-exciting dynamos which generate electric currents and magnetic fields in this way have been built in the laboratory.

One remarkable property of the Earth's magnetic field has been discovered within the past few years. In recent geological time, at intervals of a million years or so, the field has reversed—that is to say, the magnetic north and south poles have changed places. This reversal can be explained by the self-exciting dynamo theory for the origin of the magnetic field, and has profound implications in the study of the development of the Earth's crust, to be discussed in the next chapter.

Plate 10   Gold mines of South Africa are some of the deepest in the world. This chamber is nearly 2 km below the surface in the recently opened Kloof Mine, South Africa. *Photograph: Consolidated Gold Fields Ltd.*

The fact that the Moon has only a weak magnetic field suggests that it does not have a liquid metal core like that of the Earth, although some scientists think that it had such a core during its early history, when volcanic activity was taking place at its surface. Since then the Moon may have cooled sufficiently for the core to have solidified.

Internal Heat

In most deep mines and boreholes there is an increase in temperature with depth. In very deep mines this can make working conditions difficult, and special cooling may be needed. This increase of temperature with depth shows that heat is escaping by conduction from the interior of the Earth to the surface, whence it escapes into the atmosphere and outer space. The hot gas and lava poured out of active volcanoes also show that, at least locally, high temperatures are attained deep in the Earth.

We have seen, however, that the study of earthquake waves has shown that the mantle of the Earth is solid down to the Gutenberg discontinuity separating the mantle and core. If the temperature gradient which is present near the surface were maintained constant to the centre of the Earth, the melting-point of probable mantle material would be reached long before the depth of the Gutenberg discontinuity. It seems that the temperature gradient must become less in the mantle than it is in the higher parts of the crust.

What is the source of the interior heat? In the last years of the nineteenth century, when it was generally believed that the Earth originated from molten material, people thought that the inner heat of the Earth represented that part of the original heat which had not escaped. The great physicist Lord Kelvin calculated how long it would take the Earth to cool to its present condition from its original molten state. He obtained an answer which even at that time seemed to be very low. This could be explained if there were some undiscovered source of heat energy in the Earth.

That source was revealed by the discovery of radioactivity. The decay of radioactive isotopes releases nuclear energy which heats the Earth, just as the release of nuclear energy in an atomic power-station is used to heat water to generate electricity. The decay of uranium to lead, and the decay of one isotope of potassium to argon (see Chapter Ten) are especially important in supplying heat energy to the Earth. Potassium and uranium are more abundant in the crust than in the mantle, which explains the decrease in the production of heat, and therefore of the temperature gradient, in the mantle.

There are other possible sources of energy in the Earth. It may be that the solid inner core is gradually becoming larger by the solidification of liquid nickel-iron of the outer core. When the metal solidifies heat energy is released which could travel through the liquid outer core and the mantle to the surface. Heat might also be produced in the interior if the Earth were becoming smaller. On a contracting Earth the outer layers would be moving towards the centre, thus losing gravitational potential energy like a stone thrown off a cliff. This energy would be converted to heat and warm up the deeper parts of the Earth. Although the Earth is probably not contracting at present, this form of heating may have been important in the early stages of its evolution.

The quantity of heat produced in the interior of the Moon is crucial to theories of the origin of the features of the Moon's surface. If there is very little production of heat, the features must have been produced by external processes such as meteorite impact. If, on the other hand, there is production of heat on a scale comparable to that in the Earth, many surface features might be due to internal processes such as volcanic and tectonic activity, as they are on Earth. Experiments to measure the rate of flow of heat through the lunar REGOLITH were carried out by the astronauts on the Apollo 11 and 12 missions. It is not easy to know how much the temperature gradients they measured were influenced by changes in the surface temperature of the regolith, or how representative the results are of the Moon as a whole. Preliminary study of the results suggests that the flow of heat from the Moon is about the same as from the Earth, favouring the theory that the Moon has a comparatively hot interior. The study of the samples brought back by the Apollo astronauts shows that radioactive elements are present in small quantities in the surface rocks, but it is not known what their distribution is at depth. The study of samples from the lunar HIGHLAND regions, older than the MARIA where the astronauts have already landed, will be particularly interesting in shedding light on this problem.

Variations in Density

We can make another comparison between the interiors of the Earth and Moon. So far we have discussed the structure of the interior of the Earth

Plate 11 The working face of a gold mine in South Africa. The rock is composed of fragments of rocks that accumulated in stream beds many hundreds of million years ago. Although this is the richest gold mine in South Africa, one ton of rock only yields a few grammes of gold. *Photograph: Consolidated Gold Fields Ltd.*

in terms of uniform spherical shells. But the Earth is not a perfect sphere, and there are variations in the thickness and density of the different layers, especially the crust, underlying different parts of its surface. The best-known variation from a true spherical shape is the Earth's equatorial bulge, which is caused by its rotation. There are other slight departures from a spherical form, which can now be accurately determined by a study of the orbits of artificial Earth satellites.

When artificial satellites were placed into orbit round the Moon it was found that the Moon departed from a spherical shape more than the Earth. The most pronounced feature is a 'bulge' towards the Earth on the side of the Moon facing the Earth; in other words the Moon is slightly egg-shaped, with the pointed end of the egg towards the Earth. But the departures from a spherical shape are not large enough to explain the deflection of satellite orbits, so it appears that the interior layers of the Moon are also not spherically symmetrical. There are local concentrations of dense material under certain parts of the Moon's surface. Such large density contrasts do not occur in the crust or mantle of the Earth. These concentrations lie under the mare areas. They are called *mascons* and their nature is still a matter of controversy. Some scientists have suggested that they are deeply buried, large fragments of dense nickel-iron meteorites which collided with the Moon, triggering off volcanic activity to form the maria. Others think that the mare areas are not only relatively thin lava flows but that there are concentrations of relatively dense igneous rock below them.

Conclusions

Now that the evidence concerning the interiors of the Earth and Moon has been examined, it is appropriate to discuss their origins. The samples brought back from the Moon by the astronauts show that the surface rocks are quite like many terrestrial rocks, and like some kinds of meteorites, suggesting that the Earth and Moon have formed from similar parent material. On the other hand what little evidence we have about the interior of the Moon suggests that it has a different structure from the Earth, so its evolution might have been different.

Theories of the origin of the Earth and Moon are local applications of theories of the origin and evolution of the solar system as a whole. From the point of view only of the development of the internal structures they may be divided into two classes: those theories which consider that the Earth and Moon condensed to their present size from hot materials, at temperatures comparable with the melting-points of rocks; and those which consider that they condensed from cool or cold materials, at the surface temperatures of the Earth or colder.

At the end of the nineteenth century, and in the early years of this century, 'hot' theories seemed more plausible, because of the Earth's interior heat and because many surface features of the Earth were thought to be due to contraction caused by cooling through geological time. We have already described how the discovery of radioactivity provided a different interpretation for the Earth's heat, and we will see in the next chapter that the surface features of the Earth can be explained without assuming that the Earth is contracting. For these reasons, among others, 'cold' theories are currently to the fore.

It is possible to imagine an evolutionary history for the Earth and Moon which involves their separate initial growth from many bodies of the size of small asteroids and meteorites, along with smaller fragments of dust. As more material accumulated, heating of the interior would begin by radioactive decay and gravity collapse, until the melting-point of the accumulated matter was reached. The stony and iron parts would then separate (the technical term is differentiate) under gravity, the lighter silicates floating on the denser iron, like slag in a blast-furnace. In the Earth, radioactive heat has kept the core liquid to the present. In the smaller Moon, the original process of differentiation into core and mantle may not have gone so far, or the core may have formed and have solidified a long time ago. The Moon would retain the marks of the later impacts on its surface in the form of impact craters, whereas most impact sites on earth would be obliterated by erosion and tectonic activity.

Increasing knowledge of the Moon's interior derived from the lunar landing programme will probably modify our ideas about the origin of the Earth and Moon considerably, with implications for the origin of the solar system as a whole.

Chapter Four

Structure and Tectonics of the Earth

R. MASON

A physical map of the Earth, or better still a globe, reveals many features of the planet. About two-thirds of the surface is covered by sea and one-third by land. This in itself is remarkable, because we know that the land is being worn away by wind, rivers, sea and ice. If this process has gone on throughout the Earth's long history, why does the land stand above the sea at all? Why is there not a sea of uniform depth encircling the globe? Moreover, the land in places stands more than 5,000 metres above the sea. These areas are undergoing particularly rapid wearing down—as the helmets worn by climbers for protection against falling fragments of rock testify—so that the uplift to their present altitude must have occurred in quite recent times. If we look at the floor of the sea we find equally great variations in level. There are long trenches, 4 or 5 km deeper than the surrounding ocean floor. Many of these trenches are adjacent to mountainous land. Why have these irregularities not been smoothed out by the deposition of sediment on the sea-bed?

The answer to these questions is that the major features of the Earth's surface, such as seas and lands, mountains and deep ocean trenches, are not fixed things but are changing and developing. These changes are very slow on a human time-scale, but they are perceptible. The famous Pass of Thermopylae in Greece, where Leonidas and his Spartans fought the Persian army in 480 BC, on a narrow strip of land between the mountains and the sea, is now no longer a pass. The sea has retreated 2 km, leaving a coastal plain. Several villages are recorded in the Domesday Book, compiled about AD 1085, in a part of England that now lies under the North Sea.

Changes like these are not always gradual. The harbour and waterfront of Port Royal in Jamaica disappeared overnight below the sea during an earthquake in 1692. Such a rapid change in the shape of the Earth's surface always occurs in an area subject to earthquakes, and frequently where there are also active volcanoes (*Plate* 13).

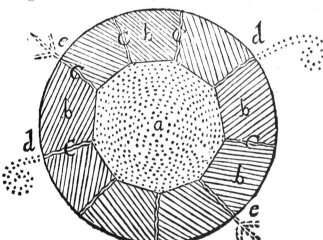

Plate 12 Older views of the Earth's interior recognised the significance of volcanoes. This author, inspired by the discovery of circulation of blood in the human body, suggested that winds circulated from the Earth's interior to the atmosphere through volcanic vents. Ideas of this type go back as far as Aristotle: (a) a mighty space; (b) crude and undigested matter; (c) cracks; (dd) hurricanes; and (ee) eruptions and earthquakes. From T. Robinson in *Anatomy of Earth* 1693.

Mobile Belts and Stable Areas, Oceans and Continents

Earthquakes do not occur with equal frequency or severity in all parts of the world. Earthquakes strong

The Earth and Its Satellite

Plate 13   An engraving of the Temple of Serapis in 1836. This area, near Naples, has changed its level with respect to the sea on more than one occasion. Here, a temple is seen to have sunk to sea-level, the pillars being surrounded by water. Sudden changes in level in this area may denote an impending volcanic eruption. From Lyell, *Principles of Geology*.

enough to cause loss of life are almost entirely restricted to a number of zones (*Figure* 10). Most of the active volcanoes of the world also occur in these zones. And, if this map is compared with a physical map of the world, it can be seen that these zones also contain a large proportion of the world's high mountains and deep ocean trenches. Thus they are zones where the surface of the Earth is undergoing particularly rapid changes. In this chapter they will be referred to as 'mobile belts', making it possible to divide the surface of the Earth into mobile belts and the stable areas lying between.

The crust of the Earth, above the MOHO DISCONTINUITY, is of two main kinds. Under continental areas the Moho lies at a depth of 30–70 km, while under oceanic areas it lies at 6–10 km below the bed of the sea. There is also a distinction in the composition of the crust between the two areas. On the continents granitic rocks form the upper part of the crust, although they are blanketed by sediments. The crust under the oceans is mostly of basaltic composition. The boundaries between areas of continental crust and areas of oceanic crust do not lie at the shores of the continents. The seas close to the shores are underlain by continental crust and constitute what is known as the continental shelf. The rocks of the continental shelf are extensions of continental rocks below the sea and are beginning to be exploited for their mineral reserves, notably of oil and natural gas. The seas of the continental shelves are shallow: less than 500 metres. The continental shelves and continents will be referred to together in this chapter as 'continental areas'. Between the continental areas and the oceanic areas is the boundary called the continental slope, where the depth of the sea increases rapidly from 500 metres to over 3,000 metres.

Mobile belts may lie in either oceanic or continental areas, or they may correspond to the boundary between the two. The Himalayan mobile belt lies entirely in the Asian continental area; that of western South America lies at the boundary between continent and ocean; while the Tonga Islands lie on a mobile belt which lies entirely within the Pacific oceanic area.

Mid-oceanic Ridges, and Ocean-floor Spreading

One type of mobile belt lies entirely in oceanic areas. This type occurs along the mid-oceanic ridges of which the Mid-Atlantic Ridge is the most striking example. Dredging of samples from the ridges shows them to be composed of a volcanic rock known as basalt, with little or no covering of sediment. Submarine volcanic eruptions occur on the ridges and their flanks, and the highest parts of the ridges stand above the sea as volcanic islands.

The ridges are zones of earthquakes, that occur in the uppermost 20 km of the crust and mantle, and are not as strong as those found in other types of

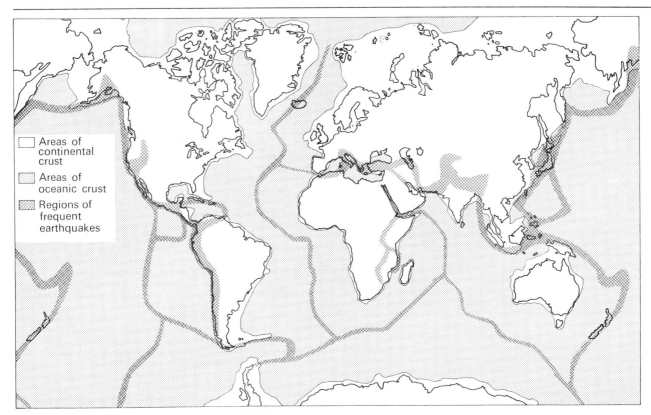

Figure 10 World map showing present-day mobile belts.

mobile belts. The earthquakes occur particularly below the narrow, trench-like valleys running along the crests of the ridges, and at points where the mid-oceanic ridges are offset by fractures in the crust and upper mantle, known as faults.

There is a remarkable pattern of variation in the strength of the Earth's MAGNETIC FIELD in the vicinity of mid-oceanic ridges. Local variations in the strength of the magnetic field are called magnetic anomalies. If the field is stronger than usual there is said to be a 'positive magnetic anomaly'; if it is weaker than usual, a 'negative magnetic anomaly'. Near mid-oceanic ridges positive and negative magnetic anomalies form stripes, parallel to the ridge on either side (Figure 12). If the ridge is offset by a fault, so are the magnetic anomalies. The stripe-like anomalies are not all of the same width, but the pattern of stripes on one side of the ridge is the mirror-image of that on the other. This pattern of magnetic anomalies was discovered about 12 years ago, and its origin was explained by two British scientists, F. J. Vine and D. H. Matthews, in 1963. To understand their interpretation it is necessary to introduce an older idea, *continental drift*, and to understand one consequence of the reversals of the Earth's magnetic field mentioned in the previous chapter.

Ever since the first maps appeared representing the east coast of South America and the west coast of Africa, perceptive people have noticed how the two appear to fit together, like pieces of a jigsaw puzzle. Could they be fragments of a larger continent which split into two or more parts in the past? This idea was developed scientifically by the German explorer and scientist Alfred Wegener, in a book published in 1915. He suggested that all continents had formed by the breaking apart of one continent, and he gave a great deal of geological and geographical (from the fitting of continental margins) evidence in support of his theory. At the time some geologists and most geophysicists disagreed with his theory, although, significantly, it was enthusiastically supported by geologists working in South America and South Africa. The chief reason for the scepticism with regard to the theory, which became known as continental drift, was that it was difficult to see how continents could move about the surface of the globe when they rest on a mantle that seismic studies show to be solid. Although some of Wegener's evidence was striking,

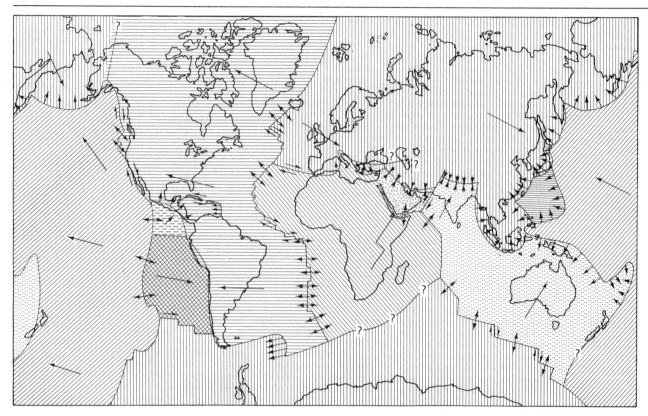

Figure 11 World map showing the larger lithospheric plates.

it was thought by geophysicists to be insufficient to support so bold a theory.

It was the study of the Earth's magnetic field which revived interest in Wegener's ideas. A molten rock is too hot to be magnetic, but when it cools it becomes slightly magnetic. This magnetism runs parallel to the magnetic field in which the rock cools. Volcanic rocks, cooling in the Earth's magnetic field, record the direction of that field at the time when they cooled. The Earth's field is inclined to the horizontal at different angles in different parts of the world, being vertical at the magnetic north and south poles, and horizontal at the equator. So the 'magnetic latitude' of any point can be found from the inclination of the field. The direction of magnetisation of a rock specimen may be found by using a sensitive instrument called a MAGNETOMETER. From this measurement the magnetic latitude of the rock at the time when it cooled may be found. If the magnetic latitudes are determined on two rock samples, which formed at the same time in two widely separated locations on one continent, the position of the magnetic poles relative to that continent can be determined.

Thus it was discovered that the continents had

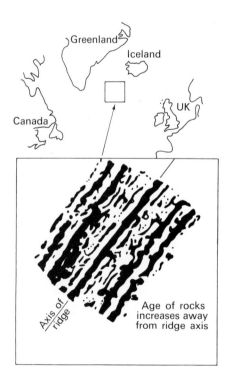

Figure 12 Map of the magnetic stripes on the mid-Atlantic ridge south-west of Iceland. Positive magnetic anomaly dark, negative anomaly light.

Plate VIII (*opposite*)
A fence offset by movement on a fault during an earthquake in western Anatolia, Turkey, on 22 July 1957. *Photograph: N. N. Ambraseys*

Plate IX
An aerial view of the volcano Surtsey, an island off the coast of Iceland. Compare with Plate X. *Photograph: J. E. Guest, February 1964.*

Plate X (*opposite*)
The active crater of Surtsey as seen from a boat 300 metres from the crater. Near the shore (left) a jet of black ash (produced by sea-water chilling the hot lava) is being erupted. Inside the crater, away from the sea, there is a fountain of fresh, hot lava. *Photograph: J. E. Guest, February 1964.*

Plate XI
An active gas vent on Mount Etna: the temperature of the gas being erupted is about 1,000 °C. Geologists on the lip of the vent are dressed in asbestos suits to protect them from the heat as they set up instruments to analyse the gas.
Photograph: J. E. Guest, October 1969.

Plate XII (*above*)
A night view of the vent shown in Plate XI. The hot gases form an incandescent cone over the vent. In the background explosions from the North-East crater can be seen (see Plate XIII) *Photograph: J. E. Guest, October 1969.*

Plate XIII (*left*)
The North-East crater on Mount Etna. Explosions from a lake of lava in the crater throw up red-hot bombs. Lava flows are erupted from the foot of this cone (see Plates XV to XVII). *Photograph: J. E. Guest, October 1969.*

Plate XIV
The North-East crater of Mount Etna at night. This photograph is a time exposure during one explosion, showing the trajectories of individual bombs. *Photograph: J. E. Guest, July 1970.*

Plate XV
Lava issuing from a small vent at the foot of the North-East crater illustrated in the last two plates. This lava has a temperature of about 1,100 °C and is very fluid.
Photograph: J. E. Guest, July 1970.

Plate XVI
A flow of ropy lava on Etna. The dark congealed flows nearby are only a few days old and still hot. *Photograph: J. E. Guest, July 1970.*

changed their positions relative to the Earth's magnetic poles during geological time. But that was not all. The continents followed different paths relative to the magnetic poles, showing that they had also moved relative to one another. Here was independent evidence in support of Wegener's theory!

All this work confirmed Wegener's conclusion that the Atlantic Ocean had come into existence in comparatively recent geological time. As this ocean grew by the separation of the continents new crust must have formed below the ocean. The most likely place for this to have occurred is at the mid-oceanic ridge, where new rock is being formed by volcanic action. This idea, an extension of Wegener's theory, is called 'ocean-floor spreading'.

The study of the magnetism of rocks also showed that the Earth's magnetic field had reversed at intervals in recent geological time as mentioned in the previous chapter. This was revealed by similar reversals in the magnetism of rocks of different ages. By dating rocks with normal and REVERSED MAGNETISM, the history of the reversals in the Earth's magnetic field during the past 30–40 million years was discovered. The reversals took place at irregular intervals, so that sometimes the field was in the same direction for several million years, while at other times it reversed after less than a million years.

We are now in possession of all the clues that Vine and Matthews used in interpreting the striped anomaly pattern across the mid-oceanic ridges. The lavas solidifying and cooling on the mid-oceanic ridges today become magnetised in the direction of the Earth's present magnetic field. The magnetism of the rocks therefore acts in the same direction as the Earth's field, thus strengthening the field over the mid-oceanic ridge, producing a positive magnetic anomaly running along the central part of the ridge. If the ocean floor is spreading from the ridge, the volcanic rocks will become older away from the ridge on each side. At a distance from the ridge, rocks will be found which formed before the last reversal of the Earth's magnetic field. The magnetism of these rocks will act in an opposite direction to the Earth's present magnetic field, thus weakening the field, producing negative magnetic anomalies flanking the central positive anomaly on either side. Yet farther out there will be rocks which formed before the last-but-one reversal, when the field was in its present direction, which will produce two positive anomalies outside the two negative anomalies, and so on.

It has been demonstrated that the different widths of the anomaly bands on either side of the mid-oceanic ridges correspond to the different intervals between magnetic reversals determined by land-based studies of rock magnetism. A wide band corresponded to a long interval, a narrow band to a short interval. This confirmed that the ocean floors are spreading from the mid-oceanic ridges, and has also enabled estimates to be made of the approximate rates of spreading from the ridges. From the width of the stripes across any particular section of mid-oceanic ridge the spreading rate can be determined. The rates are not the same on all mid-oceanic ridges, but they are all very slow on a human time-scale. The maximum rate is about 10 cm per year.

Recent studies of the ocean floors have been aimed at testing the ocean-floor spreading hypothesis. The ages of rocks from the ocean floor have been determined radiometrically and the ages of the sediments lying immediately above determined by a study of the fossils they contain. Both methods show a steady increase in age away from the mid-oceanic ridges. The rocks have also been studied magnetically to see whether their magnetism is in the direction which the Vine–Matthews theory predicts. So far, the results confirm the theory.

Mountains

We are discussing here the origin of the kind of mountain found in great chains of peaks running along mobile belts. They include most of the world's highest peaks; for example, the Himalaya, the Alps and the western American cordillera which run from Alaska to Tierra del Fuego (*Plates* 14, II, III). In all mountains of this kind there is evidence for great uplifts of the surface of the Earth. Leonardo da Vinci noticed the remains of marine animals in the rocks high in the Apennine mountains of Italy, and fossil marine organisms have been found at 8,500 metres on Mount Everest. There is not enough water in the world, even if the polar ice-caps were melted, for the sea to rise to such a height. Thus in the mountain areas large parts of the Earth's surface must have been raised several kilometres relative to sea level. It is possible that this process of uplift is still going on in some mountain chains. For example, the measured height of Mount

Plate 14 Part of the Andean mountain chain between Chile and Argentina. Mount Aconcagua is the high pyramidal mountain on the skyline. Near-vertical rock strata are seen in the valley side to the left of the picture. The whole area was glaciated during the Pleistocene to give the deep valleys; now only remnants of glaciers remain high on the mountain slopes. *Photograph: J. E. Guest.*

Plate 15 Earthquake damage in Chimbote, Peru, after an earthquake on 31 May 1970. Note how thick brick walls have collapsed, while light wood walls and telegraph poles, which can vibrate freely with the ground, remain standing. *Associated Press photograph.*

Everest has increased in each successive survey. This may be coincidence, as the early surveys were not so accurate as the modern ones and the increases are only a matter of a few metres, but there is independent evidence of rapid uplift of the Himalaya in the recent past and it would not be surprising if this were continuing.

There is also evidence for subsidence of the Earth's surface in regions that are now mountain ranges. When sedimentary rocks of a particular age are traced into a mountain chain from the adjacent stable area they frequently show an increase in thickness towards the heart of the mountain chain. The sediments forming the mountains are sometimes of a type which it is known can form only in shallow water (see Chapter Six), although they may be as much as 10,000 metres thick. This can be explained only if the sediments formed in a shallow sea whose bed was steadily sinking, so that the infilling of the sea by sediment kept up with the sinking of the sea-floor (*Figure* 22). The name given by geologists to such a subsiding region is a *geosyncline*.

There are many indications that mountain ranges are regions where the rocks have been deformed due to *shortening* of the Earth's crust. The originally horizontal layers of sedimentary rocks are often crumpled (*Plates* 17, IV). This can be seen on all scales from crumples tens of kilometres across to single crumpled crystals less than a millimetre across. The crumples, called *folds* by geologists, are

Plate 16 An aerial photograph showing mud walls round fields offset during an earthquake in Iran. The fault complex along which movements occur is marked by an arrow. *Photograph: N. N. Ambraseys.*

Plate 17 The layers of rock in this photograph were deposited horizontally on the sea-bed. They are now folded into a steep anticline. The small arrows mark the top of one layer that was once horizontal. Balkan range, Bulgaria. *Photograph: D. T. Donovan.*

not randomly arranged. Their axes show a tendency to be parallel to the direction of the mountain chain. The wearing down of the mountain range emphasises this feature, so the harder bands of rock stand out as ridges parallel to the mountain chain.

Shortening may also be indicated by deformation of objects whose original shape we know, such as fossils (*Plate* 18) or the spherical coloured patches which often occur in soft mud. In the deeper parts of the mountain belt heating can occur during shortening, so that the rocks recrystallise. These rocks are said to have been METAMORPHOSED, and the orientation of the recrystallised minerals often indicates the direction and intensity of shortening. The greatest shortening usually occurs at right angles to the direction of the mountain chain, but the details of the deformation are complex, and shortening can occur locally in any direction.

In some mountain chains shortening becomes so intense that rocks not only crumple and change shape, they break along nearly horizontal fractures, so that one block of rocks overrides another. Sometimes sequences of rock several hundred metres thick are turned upside-down, so that younger rocks are overlain by older ones, and rocks of the same age may reappear two or three times up a mountainside. Structures of this sort are well known in the western Alps of France and Switzerland, which appear to have been regions of particularly intense structural shortening.

Geophysical studies in mountain areas reveal that mountains have 'roots', that is to say there is an increased thickness of the crustal layer of the Earth. Studies of earthquakes show that the Moho discontinuity between the crust and mantle is at depths of 60–70 km under the Alps and many other mountain chains, compared with about 30 km under the surrounding stable regions. The presence of a greater thickness of less dense crustal rocks in a mountain chain causes the Earth's gravity field to be locally weaker.

Geologists have worked out the typical history of a mountain chain from a study of many mountainous regions. Subsidence of the Earth's surface comes first, sometimes associated with submarine volcanic activity. When a thick sequence of sediments and possibly volcanic rocks has accumulated in the geosyncline, intense shortening of the Earth's crust causes crumpling, distortion and fracturing of the rocks. Finally the rocks are uplifted to form a

The Earth and Its Satellite

Plate 18 This fossil trilobite was once bilaterally symmetrical; its left side was a mirror image of the right. Mountain-building movements have distorted it to its present shape. *Photograph: University College London, Geology Department Photographic Collection.*

mountain chain, and this process may also be accompanied by volcanic activity. In any particular mountain chain the details of the history are more complicated than this. The geosynclinal phase usually involves the development of several geosynclines, some with accompanying volcanism, some without. Several phases of shortening in different directions may occur during the phase of folding.

The structures formed by the mountain-building process are so characteristic that it is possible to recognise ancient mountain chains which mark the sites of mobile belts which were active at different periods throughout geological history, but which are now part of the stable continental areas. In fact, the continental areas are formed almost entirely of ancient mobile belts, some now buried under more recent sediments.

Island Arcs, Mobile Continental Margins, and Disappearing Ocean Floor

The deepest and most intense earthquakes occur in the mobile belt surrounding the Pacific Ocean and along those parts of the Alpine-Himalayan belt that lie along continental margins such as the East Indies and the Mediterranean region. On the eastern side of the Pacific a number of islands occur on arc-shaped ridges, often with an arc-shaped deep ocean trench on the oceanward side. These islands are volcanic, but some rest on continental crust (e.g. the Japanese islands, Java, Sumatra) while others rest on oceanic crust (e.g. the Tonga Islands, Marianas Islands, New Hebrides). Accurate location of the foci of the earthquakes shows that the shallowest occur on the landward side of the deep trench, and the deeper earthquakes lie on a surface sloping at about 45° under the island arc, away from the trench, going down as far as 700 km. These surfaces are named 'Benioff seismic zones' after the seismologist who discovered them (*Figure* 13). By studying the way earthquake waves diminish in strength away from the foci on the Benioff zones, seismologists have shown that there is a zone of rock which is unusually rigid and dense lying immediately beneath the Benioff zones.

To explain what may be happening under ISLAND ARCS we must say more about the structure of the upper mantle. There is a layer in the mantle at a depth of between 100 and 400 km which has different properties from the layers above and below. *P* and *S* earthquake waves travel more slowly in this layer. This means that they are deflected into it from above and below, so that it acts as a 'wave guide', transmitting a relatively large proportion of the earthquake energy. Waves tend to diminish in strength more quickly in this layer than in the mantle above and below. It has become known as the 'low-velocity layer'. It is not bounded above and below by sharp discontinuities, like the mantle itself, but passes gradually into more rigid mantle on either side. It is still not known whether the low-velocity layer is present in the mantle right round the world, but its presence has been detected on either side of the Benioff zones.

One further development in seismology has influenced our knowledge of the earthquakes beneath island arcs. By noticing the direction of the first displacement of the rocks by the *P* waves arriving at a number of different seismograph

Structure and Tectonics of the Earth

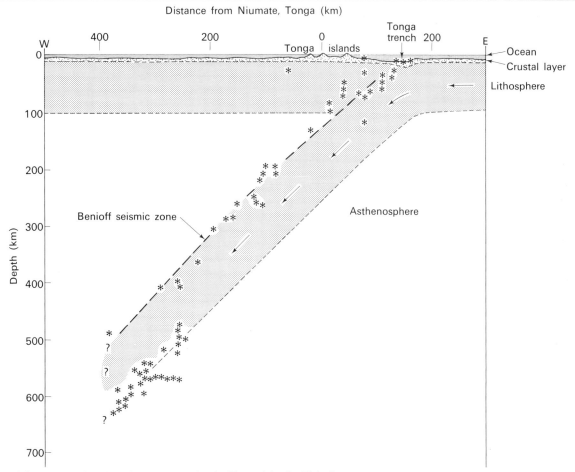

Figure 13 The pattern of earthquake activity under the Tonga islands. This diagram is true scale. Small stars represent foci of earthquakes. The cause of deep earthquakes is not known, but they may be associated with the break-up of the descending lithospheric plate. The thickness of the crust and lithosphere are approximate.

stations, the direction of the stresses in the crust and mantle which caused the earthquake can be calculated. This kind of study confirms that earthquakes near mid-oceanic ridges are caused by tension in the spreading crust. Under island arcs the shallow earthquakes, down to 100 km or so, are caused by the block of rigid rock sliding below the Benioff zone against the rocks above. The deeper earthquakes do not conform to this pattern, for reasons which are not understood.

From this evidence it appears that the rocks below the Benioff zone are the crust and upper mantle, being forced beneath the island arc down into the low-velocity layer of the mantle. Thus, new crust is being created at the mid-oceanic ridges, and is disappearing down below the Benioff zones. Benioff seismic zones occur under some continental boundaries (e.g. the west coast of South America) and it appears that here, too, oceanic crust is disappearing into the mantle. However, nowhere is continental crust disappearing below a Benioff zone.

Transcurrent Faults and Transform Faults

In some areas, such as the San Francisco region of California, earthquake activity is associated with great fractures in the Earth's crust, known as faults. After the disastrous San Francisco earthquake of 1906 the rocks on either side of one such fault, the San Andreas fault, were displaced by as much as 6 metres. This movement was not in a vertical direction but a horizontal one, so that fences and roads which had been continuous across the fault were offset to the right by the earthquake. A study of the landforms of the region shows that streams in narrow canyons are offset by as much as 50 metres, also to the right, probably as the cumulative result of a number of smaller movements like that associated with the San Francisco earthquake. Still greater offsets, of several kilometres, can be demonstrated by comparing rock-types on either side of the fault. All the offsets are in the same direction. It appears that the floor of the Pacific Ocean and a narrow strip of the coast of California are sliding north relative to the main part of North America, and that the junction is the line of the San Andreas fault. Faults of this kind, along which the displacements are almost horizontal, are called 'transcurrent faults'.

Transcurrent faults occur in other parts of the world also. A British example is the fault which runs along the Great Glen of Scotland, separating the northern Highlands from the Grampian Highlands. An offset of 100 km has been demonstrated by a comparison of the rocks on either side of the fault, but in this case the offset is to the left, not the right as on the San Andreas fault.

The faults which offset the mid-oceanic ridges have already been mentioned. These faults are different from transcurrent faults because of the spreading of the Earth's crust which is taking place along the ridge. This means that there is relative movement on either side of the fault only on the short section of the fault that lies between the two offset parts of the ridge. In this section, the relative movement of the rocks on either side of the fault is in the *opposite* direction to the offset of the ridge. These faults were given the name 'transform faults' by J. T. Wilson, who first realised that they were different from transcurrent faults, and first motion studies of earthquakes occurring on them shows that the relative motion of the two sides of the faults is in the direction he predicted.

A Theory of the Movements of the Earth's Crust— Plate Tectonics

If the Earth is not expanding or contracting, the rate of production of new crust at the mid-oceanic ridges must balance the rate of disappearance of crust at island arcs and mountain chains. This balance is the central idea in the modern theory of the development of the Earth's surface known as 'plate tectonics'. According to this theory the mobile belts are the junctions between rigid plates composed of the crust and the uppermost 100 km or so of the mantle. These plates are in motion relative to one another, moving on relatively plastic rocks in the low-velocity layer of the mantle below.

The crust and upper part of the mantle which behave as one mechanical unit are called *lithosphere* and the low-velocity layer is called *asthenosphere*. Large-scale movements in the mantle and crust occur only at the boundaries between the lithospheric plates, and the continents drift passively in the upper part of the lithosphere, like logs frozen in the ice of a lake.

Where two plates are moving apart there is a mid-oceanic ridge, with magma welling up from the asthenosphere to be erupted as basalt and form a new crust, while the unmelted material remains to constitute the lower part of the lithosphere. At island arcs and tectonic continental margins the whole lithosphere descends below the Benioff zones into the asthenosphere, while in mountain ranges the crust shortens by crumpling and fracturing. At transform and transcurrent faults one lithosphere plate passes another with no loss or gain of crustal material (*Figure* 14). Since lithosphere is being lost into the mantle below Benioff zones and appearing at mid-oceanic ridges, there must be a counter-flow of material in the mantle. This is thought to occur in the asthenosphere.

The character of mobile belts at the present day can be explained in this way by the movements of six major lithospheric plates, as shown in the map (*Figure* 11). It should be remembered that the plates are moving over the surface of a sphere, so that their movements are rather difficult to represent on a flat map of this kind.

The Stable Regions, and Isostasy

Although the stable areas of the lithospheric plates do not undergo large horizontal deformations, some movement does occur. In these regions, vertical movements of the Earth's surface predominate and are slower than the kind of uplifts and downwarps which occur in mobile belts. Large areas of the stable continental areas may be downwarped and flooded by the sea. Hudson's Bay, the Baltic Sea and the North Sea are present-day examples of this. Other areas may be uplifted into ranges of hills or even mountains, such as the Appalachian mountains of the eastern USA or the mountains of Norway and Scotland. Although these areas are the sites of ancient mobile belts, their present uplift occurred long after these areas had become stabilised.

We do not understand all the causes of move-

Structure and Tectonics of the Earth

ments of this type, but one reason for vertical movements in the Earth's surface has been known for a long time. The stable parts of the lithospheric plates are in hydrostatic equilibrium, with the lithosphere floating on the asthenosphere. Thus the stable ocean floors, which have a thin, dense crust overlying dense upper mantle, lie about 2 km below sea-level, while continental areas such as much of Africa, where the crust is unusually thick, are plateaux raised up to 1,500 metres above sea level. If the weight of the crust is increased, for example by the deposition of sediment, that area will become depressed. If the weight decreases, for example by erosion, that area will be elevated. A good example of the latter process is seen in Scandinavia. This part of Europe was covered by a thick ice-cap during the last phase of the ice Age. Now that the great weight of ice has melted, the land round the Baltic Sea is rising at about 1 cm per year. Movements of this kind are called 'isostatic adjustments'. The rise must be accompanied by flow of material into the area below the Baltic Sea, and it is thought that such a flow takes place in the asthenosphere.

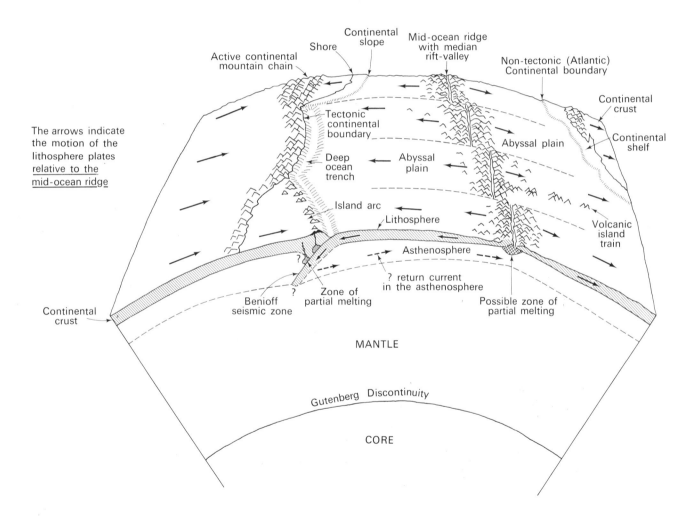

Figure 14 Hypothetical diagram illustrating the concepts of plate tectonics.

Chapter Five

Magmatic and Volcanic Processes on Earth

M. K. WELLS

The Nature of Magma and Igneous Activity

MAGMAS, the molten substances from which igneous rocks are derived, have been formed by localised melting of the mantle and/or deeper parts of the crust. There has been much speculation concerning the causes of such melting, and the following have been considered as possibilities: accumulation of heat from the decay of radioactive elements; heating from friction caused by movement of segments of crust and mantle along major dislocations; lowering of melting temperatures through the accession of water carried down to great depths. Whatever the causes, no one can doubt their effectiveness. Some form of magma 'reservoir' (the nature of which it is best not to try to define too closely in the present state of knowledge!) lies beneath every volcanic region. Each of these has a long history, spread generally over millions or tens of millions of years. There is abundant evidence that similar igneous activity has affected different parts of the Earth throughout geological time.

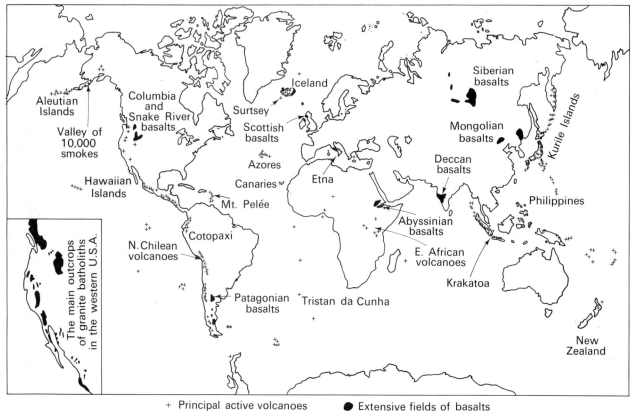

Figure 15 World map showing distribution of active volcanoes. Large fields of basalts are shown in black.

Plate 19 Ropy lava. Etna. *Photograph: M. K. Wells.*

In a sense, this igneous activity has been responsible for the creation of new crust beneath the oceans and has provided all the raw materials of which continents are built.

Since magmas are less dense than rocks of their own composition they tend to rise from the depths at which they are generated to levels where the weight of the 'column' of magma is balanced by that of the surrounding rocks, rather like mercury in a barometer. The behaviour of magma is governed by many factors, such as levels of original melting, chemical and physical properties of the magma, and stress conditions of the overlying rocks. Magmas penetrate most easily and rise fastest in the fractures developing in zones of crustal tension.

During a long history of igneous activity in such an environment magmas may in turn penetrate large numbers of roughly paralleled fractures in the crust to form what is known as a dyke swarm. DYKES are the commonest kinds of igneous intrusions. They are generally vertical, parallel-sided sheets of rock, which may extend laterally for several kilometres, though their widths are generally of the order of 1 or 2 metres (*Figure* 16).

Magmas inevitably undergo changes in composition and physical properties during their ascent. They are subjected to reductions of pressure, and they lose heat by contact with the progressively colder rocks they penetrate. Reaction with these

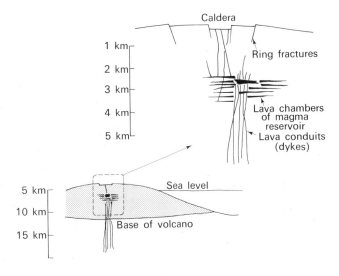

Figure 16 Cross-sections through an Hawaiian volcano showing the caldera pit and the feeder dykes.

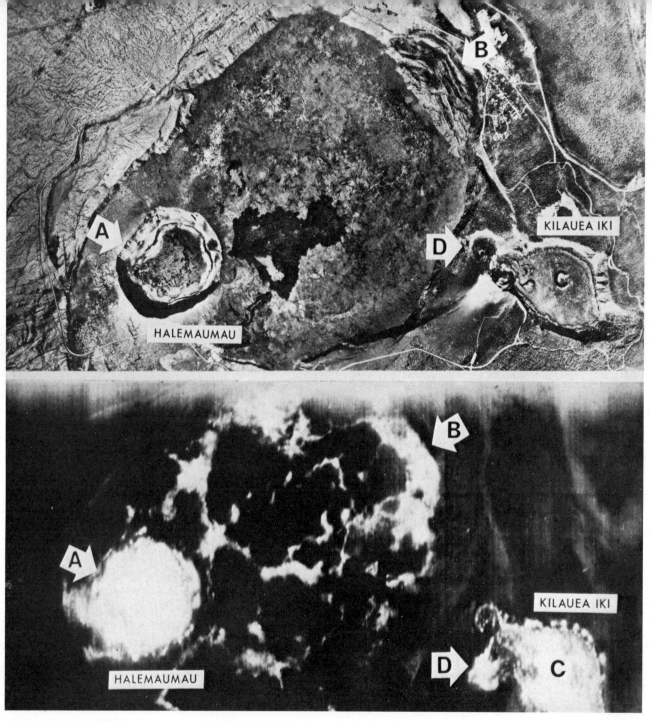

Plate 20 Kilauea caldera, Hawaii. The upper photograph shows it as seen with an ordinary camera. The deep pit at (A) is known as Halemaumau (see plate 21). The lower view of the same area is with an infra-red detector; the bright areas are those that are hot. *Photographs supplied by US Embassy, London.*

rocks also causes changes in composition as some components of the rocks may melt and become incorporated in the magma, and some of the magma's own components are precipitated.

Under conditions of crustal tension where magmas have a relatively easy passage to the surface, these changes may be kept to a minimum, and consequently the volcanic products show little variation; but in a mountain belt where the crust is thickened and broadly speaking under compression, the situation is very different. Here the magmas are subject generally to much wider variations than those of, say, a mid-oceanic rift zone, as shown by the greater diversity of their volcanic products.

The evidence of igneous activity in mountain chains is not confined to surface volcanics: erosion has bitten deep into the crust, exposing not only older volcanic rocks but also great volumes of intrusive rocks consolidated from magmas that never reached the surface. These deep-seated or plutonic

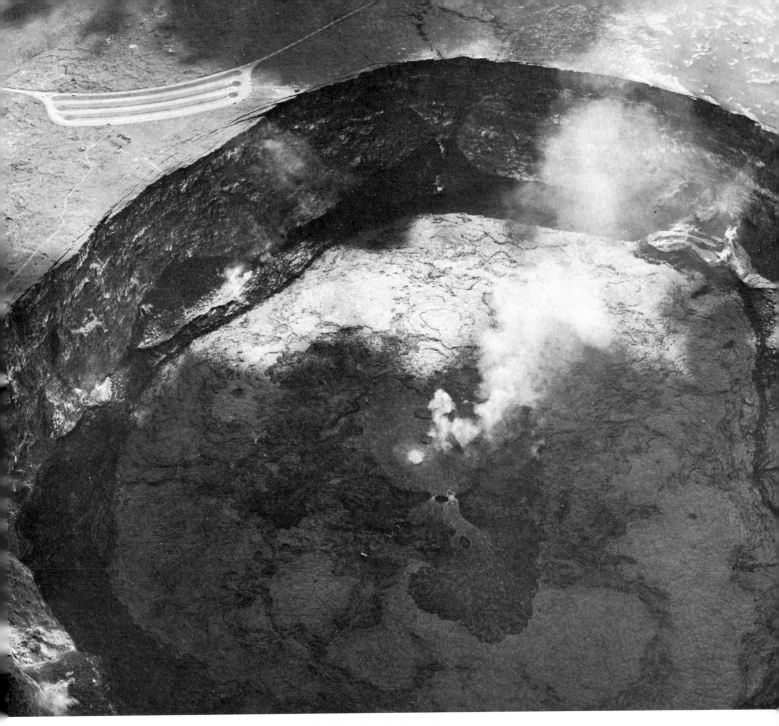

Plate 21 An aerial view of Halemaumau (see Plate 20) during the 1967–8 eruption. A small active lake has formed in the floor. *Photograph: R. S. Fiske, US Geological Survey.*

rock bodies are rightly called major intrusions because many of them have outcrops measured in tens of kilometres and they extend downwards to great, and generally unmeasured, depths.

The Rôle of Volcanic Gases

Magmas are essentially silicate melts. In addition, however, they always contain small amounts of water and other volatile components. Although the latter amount to only 1% or 2% by weight of the total, they exercise a disproportionately great influence on nearly every aspect of igneous activity. At depth the high pressures keep them dissolved in the magma. They drastically lower melting temperatures. A water-free mixture of the component minerals of granite, for instance, may melt at about 1,000 °C but, with a few per cent of water, melting may occur at temperatures as low as about 700 °C. The volatiles also act as fluxing agents, giving

Plate 22 (*above*)   Halemaumau pit (see Plate 21) from the edge of the crater, during the 1967–8 eruption. Flows from the lava lake show typical ropy structure. *Photograph: R. S. Fiske, US Geological Survey.*

Plate 23 (*below*)   Close-up view of one flow similar to those in Plate 22. *Photograph: R. S. Fiske, US Geological Survey.*

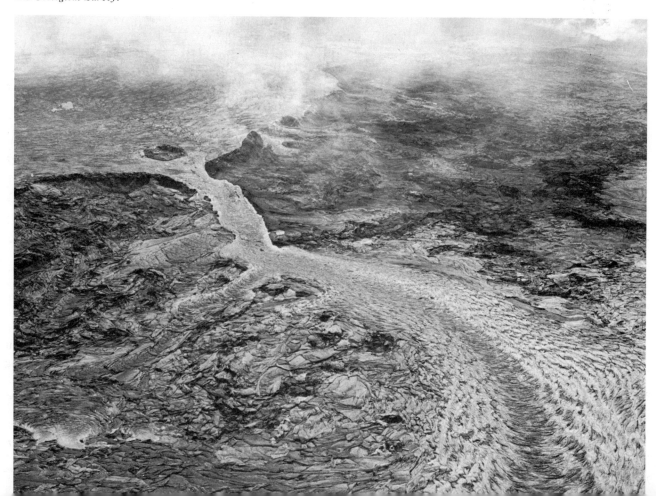

Plate 24 (above) A night view of the Halemaumau pit during the 1967–8 eruption. *Photograph: R. S. Fiske, US Geological Survey.*

Plate 25 (below) Night view of the lava lake during the 1967–8 eruption. Lava fountains 'boil' at the surface. *Photograph: R. S. Fiske, US Geological Survey.*

The Earth and Its Satellite

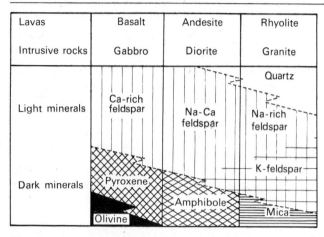

Figure 17 The principal igneous rock types and their constituent minerals.

components the freedom of movement necessary to build relatively large crystals. This factor, combined with slower cooling, makes deep-seated igneous rocks more coarsely crystalline than their extrusive equivalents.

Various factors cause the volatile components to become concentrated towards the top of magma reservoirs beneath volcanoes. The extent to which this concentration causes gases to be evolved depends on the strength of the overburden: i.e. on how well the lid of nature's pressure-cooker has been screwed down. Volcanic cones generally include enough porous and fissured rock to provide a kind of safety-valve, with some gases escaping continuously (*Plate* XI) even if there is no other kind of activity. Eruptions of fluid magma allow most of the gas to boil away relatively gently from lava surfaces. At the other extreme, viscous magma and the ASH and fragmented rock which frequently accompany its eruption may completely choke the outlet of a volcano and lead to a big build-up of pressure from volatiles accumulating beneath.

Apart from playing an important mechanical role in volcanic eruptions gases are also chemically very reactive, being composed mainly of steam, with substantial quantities of carbon dioxide and sulphur gases, and smaller amounts of chlorine, fluorine, etc.

Chemical and Mineralogical Composition of Igneous Rocks

The SILICATE MINERALS which comprise over 95% of the components of most igneous rocks are compounds of oxygen and silicon together with other elements, mainly aluminium, iron, magnesium, calcium, sodium and potassium. The stability of silicates in the crust and mantle is due to the strong affinity that exists between silicon and oxygen atoms. Atoms, or more strictly their electrically charged equivalents termed IONS, can be envisaged as exceedingly minute spheres of matter, each kind of ion having a specific radius and positive or negative charge. The positively charged silicon ion has just the right size to fit snugly between four larger and negatively charged oxygen ions when the latter are in contact with one another. This combination is the fundamental structural unit of the greater part of Earth matter and deserves a special name: mineralogists call it the SiO_4 tetrahedron.

Affinity between silicon and oxygen extends into the molten state and even here SiO_4 units tend to link up into larger structures which differ from crystal structures mainly in being less extensive and geometrically less perfect. Also, of course, in a melt the pattern of linkages is continuously changing. However, in a melt rich in silicon the changes are sluggish and the linkages tenacious. Such melts are therefore very viscous unless a sufficiency of water is present in solution to break the linkages. This means that very siliceous magmas—those of granitic composition—stiffen rapidly when pressure is released and volatiles escape during eruption. This is an explosive situation, and although the melt normally congeals into a non-crystalline glass it is only rarely able to flow as a lava, and generally becomes blown into fragments by expanding gases.

The tendency for SiO_4 to link up into more extensive structures in the melt is reduced when less silicon is present and when other elements, particularly iron, are abundant. Silicon-poor magmas,

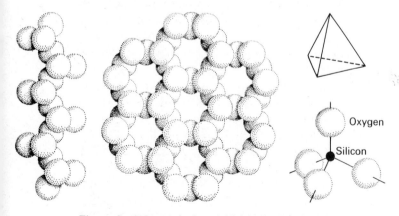

Figure 18 SiO_4 tetrahedron (*right*) indicating the relative sizes of silicon and oxygen ions. In the chain structures (*left*), the fundamental structure of pyroxenes, and the sheet structure (*middle*) characteristic of mica, only the oxygen ions are shown.

like those of basaltic composition, are therefore more fluid than highly siliceous ones, even under surface conditions. It is easy to understand why the most extensive and widely distributed lavas are basalts.

The character of volcanic activity is thus very dependent on the compositions of magma which, as one might suppose, vary widely. Since any one magma can give rise to a number of different rock types (e.g. coarse-grained intrusives, fine-grained and glassy lavas or extrusive fragmented rocks) the range of igneous rocks is even greater. Fortunately the majority of igneous rocks fall into three main categories which can be termed basaltic, andesitic and granitic (*Figure* 17).

As stated above, basalt is the rock which forms most of the world's lavas. It is almost the only type to be erupted in the oceanic areas and the crust beneath the oceans is essentially basaltic. Basalt also occurs extensively on the continents. Andesite is virtually confined to mountain chains and island arcs where lavas and ashes of this composition commonly exceed basalts in amount. Granite is king of the deep-seated and coarse-grained rocks found in the enormous batholiths intruded into the roots of mountain chains, and it also occurs extensively among the metamorphic rocks of shield areas and in other continental environments.

Some characteristics of these rocks and the types of volcanic activity with which they are associated are summarised in the remainder of this chapter.

Basalts and Basaltic Volcanicity

Basalts (*Plate* XLV) are minutely crystalline rocks, commonly composed of about equal amounts of felspar and dark coloured minerals which include a few per cent of the magnetic iron-oxide mineral, magnetite. This composition accounts for the dark colour of basalts and their density (about 2·9–3·0), which is greater than most other crustal rocks.

Basalts can often be recognised from the presence of small rounded cavities, or vesicles, formed as gas bubbles trapped in the lava during consolidation. The bubbles collect towards the lava surface which may be coated with rough-surfaced blocks of so-called scoriaceous lava (*Plate* XVII). These look just like big lumps of clinker or coke: an appropriate analogy since they are byproducts of gas manufacture! In most cases, however, even if such blocky lava is present much of the lava surface is typically formed of dense basalt with the congealed skin puckered and drawn out into lobes and festooned patterns by flow. This is appropriately termed ropy lava (*Plates* 19, XVI). Low viscosity allows basaltic lava to flow long distances and at considerable speeds even when the gradient is low. Lava may issue from a crater rim or from fissures on the flank of a volcano and will flow down any valley in its path, parting on either side of some obstructions and cascading over others like a river—at night, a river of fire—which deepens and widens as it descends.

Repeated outflow of basaltic lava from a central source builds up a so-called shield volcano, typified by the giant example of Mauna Loa in Hawaii, with a broad base and gently sloping sides. Etna approaches to this form.

In past geological times enormous areas of the crust have been flooded by 'plateau-basalts'. As the name implies, the flows are sensibly flat sheets, each extending perhaps over several hundred square kilometres. The surface may be porous and easily weathered, but the main part of the flow is generally massive and gives evidence of contraction on cooling to form a polygonal pattern of columnar jointing (*Plate* 26). Erosion of alternately massive and weathered lava tends to give a terraced topography. In some areas the total thickness of the lava sequence may be 1,000 or 2,000 metres.

Despite the impressive scale of plateau basalt occurrences, very few eruptions of this kind have been observed in historical times. The best-known example occurred at Laki in the central rift-zone of Iceland in 1783. A great sheet of lava welled out on either side of a 30 km fissure. The line of the fissure is marked by a series of small cinder cones, and one can envisage that beneath the surface there is a feeder dyke or series of dykes adding to the dyke-swarm which is associated with any region of plateau basalts.

The Laki situation is particularly instructive because, as explained above, Iceland provides a sample of conditions on the Mid-Atlantic Ridge. It seems probable that much of the submarine volcanicity is of fissure-eruption type, presumably on a scale which exceeds anything on the continents. Plates 28, 29, IX, X show the activity of Surtsey, off Iceland. This volcano started as a submarine volcano on the sea-floor, gradually building up until its top rose above sea level.

Much of our knowledge of the mechanisms of

basaltic volcanoes has resulted from studies on Hawaii (*Plates* 20–25). Here for many years a team of scientists has kept a comprehensive record of the various aspects of volcanic activity, like doctors attending some sleeping giant in an intensive-care unit, continuously recording temperature, blood pressure and pulse rate. By seismic records they have been able to trace the ascent of magma from its first detectable movements at depths of 50–60 km, generally several months before eruption. As magma invades chambers only a few kilometres below the surface the volcano summit gradually swells up, only to collapse again—much more rapidly—during eruption. Although these volume changes may amount to thousands of cubic metres, they are imperceptible without careful measurement.

Hawaiian experience shows that the source of basalt magma lies well within the upper mantle, where evidence suggests that the rocks are peridotites composed mainly of olivine. Melting experiments on peridotites carried out at pressures appropriate to mantle conditions have yielded initial fractions of liquid of basaltic composition. It seems likely that this process, known as selective fusion, is in essence the means by which new basaltic crust is created from mantle rocks in the mid-oceanic regions.

Andesites: Volcanicity of Mountain Chains and Island Arcs

The very name andesite suggests its association with mountain belts. Predominantly andesitic volcanoes are characteristic of the young mountain chains of North and South America and they are abundant in most of the island arcs including the Aleutians, Kuriles, Japan, Philippines, Java and Sumatra. This great zone of volcanic activity forms a veritable 'ring of fire of the Pacific'. Basalts are commonly present, and in some regions volcanoes of granitic composition—into which the andesitic rocks grade with increasing silicon—are particularly

Plate 26 Columnar jointed basalt of the Giant's Causeway, Northern Ireland.
Photograph: D. Savage.

Plate XVII
Scoriaceous lava flowing high on the flanks of Etna. *Photograph: J. E. Guest, July 1970.*

Plate XVIII
The nearly 6,000-metre-high volcano of Cotopaxi in Ecuador. Although this volcano has been active in historical times, its crater is now filled with ice. The crater is about 1 km across. *Photograph: J. E. Guest.*

Plate XIX *Nuées ardentes* rushing at more than 100 km per hour down the slopes of Mayon volcano in the Philippines. Clouds of this type consist of gas charged with small fragments of hot lava (ash), and they destroy everything in their path. *Photograph: J. G. Moore, US Geological Survey, May 1968.*

Plate XX (*overleaf*)
Boiling water from a geyser at Tatio in northern Chile. The intricately patterned surface consists of chemical precipitates from the water. This material is known as sinter. The red coloration is due to algal growths near the hot water. *Photograph: J. E. Guest.*

Plate XXI (*on preceding page*)
A chimney of sinter built up over a geyser at Tatio, northern
Chile. The form of this material resembles stalagmite in caves.
Photograph: J. E. Guest.

Plate XXII (*above*)
A Gemini 4 photograph of the Richat structure in Mauritania,
Africa. The origin of the structure, which is nearly 50 km across,
is not known. According to some scientists it is the root of an
impact crater similar to those on the Moon. However, it may
well be a dome caused by the intrusion of lava below the surface.
NASA photograph.

Plate XXIII (*opposite*)
The Sahara desert in North Africa. The large pale-coloured basin at upper right is
the 400 km wide Marzurq Sand Sea. *NASA Gemini 11 photograph.*

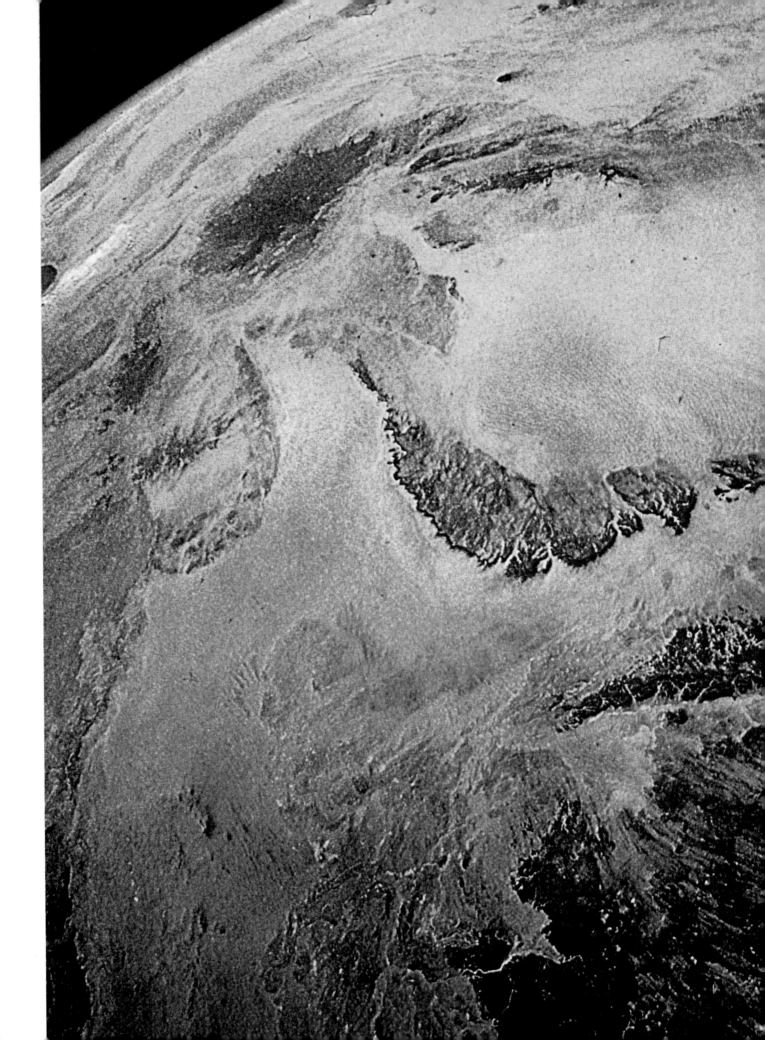

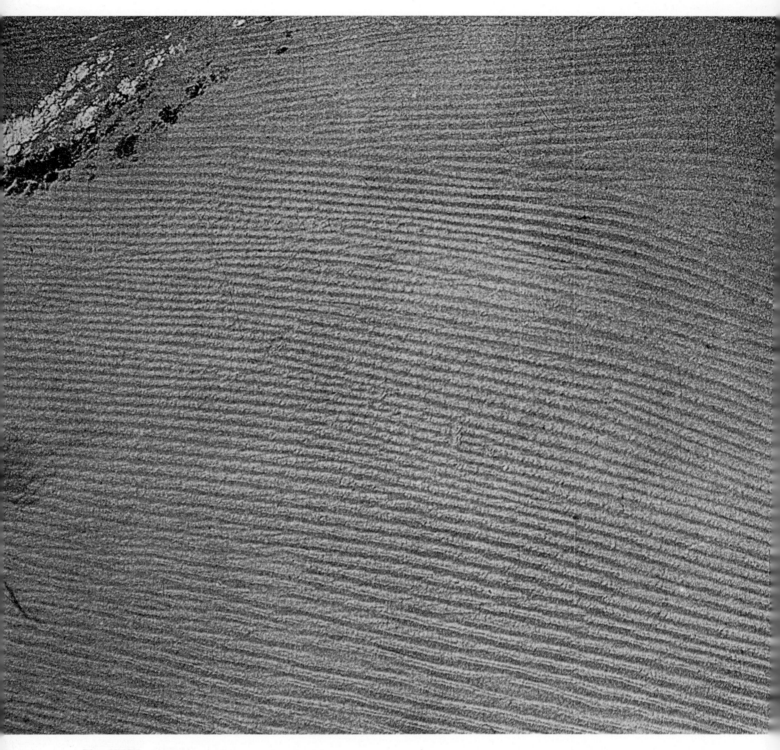

Plate XXIV
The Empty Quarter in southern Arabia, showing several thousand square kilometres of sand dunes. Some of these dunes extend for nearly 200 km without a break.
NASA Gemini 4 photograph.

abundant, though generally andesites are predominant. Both andesitic and granitic rocks are virtually confined to regions with continental crust.

Andesites contain more silicon and less iron and magnesium than basalts (see *Table* 3), and of course, this is reflected in the mineral content and properties of the rocks. They generally contain more felspar than basalts, with perhaps about 30% of coloured minerals of fairly similar kinds to those of basalts, though olivine is absent. Typical andesite is thus lighter in colour and in weight than basalt.

TABLE 3

	Basalt (Hawaii) Weight %	Andesite (Andes) Weight %	Granite (Arran, Scotland) Weight %
SiO_2	42·30	58·50	75·65
Al_2O_3	10·50	18·10	11·89
TiO_2	2·41	0·99	0·28
FeO	14·00	6·50	2·00
MgO	14·90	2·81	0·15
CaO	12·80	6·88	0·91
Na_2O	1·56	3·90	3·44
K_2O	0·42	1·95	4·26
MnO	0·06	0·09	0·26
P_2O_5	0·33	0·21	0·16
H_2O	1·32	0·30	0·81
Total	100·60	100·23	99·81

Chemical analyses of typical terrestrial igneous rocks. Note the wide differences in SiO_2 content, as well as those of FeO, MgO, CaO and K_2O. Water (H_2O) is present in all these terrestrial rocks; compare these analyses with those in Table 4 (Chapter Nine) of lunar lavas which have no water.

Because of its viscosity, andesitic magma forms restricted lava flows, and eruptions are frequently explosive, often giving rise to as much volcanic ash as lava. Although fine dust may be carried far from its source, the main concentration of ash—and particularly larger fragments—builds up close to the crater. This leads to development of the shape normally associated with a volcano (*Plates* 32, XVIII), that of a steep-sided cone such as the beautifully symmetrical example of Fujisan (Fujiyama) in Japan.

Granites and their Volcanic Equivalents

Most readers, perhaps unintentionally, are familiar with granite because of its widespread use as build-

Plate 27 The early stages of the eruption of Surtsey, Iceland. Sea-water entering the crater through the gap in the rim is causing jets of black ash and steam to be ejected hundreds of metres in the air. *Photograph: J. E. Guest, November 1963.*

ing stone. Among the best-known British examples are the rather dark Shap granite from the Lake District, containing large and conspicuous pink felspars, and granites from South-west England which are paler and contain even bigger, white felspars

Plate 28 As Plate 27. The island is surrounded by a cloud of steam and ash. Falling ash darkens the sky to the right. *Photograph: J. E. Guest, November* 1963.

Plate 29 Late in the development of Surtsey, sea-water was prevented from entering the crater and fountains of red-hot lava were thrown up. This activity was a prelude to the eruption of a large lava flow. *Photograph: J. E. Guest, February* 1964.

commonly up to 5 cm long, with rectangular shapes and a pearly lustre due to light reflected from cleavage surfaces. Dull grey translucent quartz grains lie between the felspars. About a third of the mineral content is quartz in many granites. There is generally only a small percentage of coloured minerals, dark brown or black mica being the most distinctive.

Comparison of granites with their volcanic equivalents makes it hard to realise that they can be derived from the same kind of magma. Factors affecting such a magma at or near the surface have already been discussed in general terms. They are, first, the development of high viscosities and, second, the build-up of high vapour pressures as volatiles become concentrated beneath the volcano. As a result the products of eruption are largely glassy rocks such as obsidian, pitchstone and pumice, which in the main become fragmented to form various kinds of ash accumulation. The latter include ash flows, or ignimbrites, discussed further below. True lava flows of obsidian or pitchstone are very restricted due to the rapid increase in viscosity of the melts as volatiles escape. This is illustrated clearly by the nature of pumice, in which solidification is so rapid that what one would expect to be the transient shapes of expanding gas bubbles are frozen in a kind of stony froth.

Quite commonly the last act in a cycle of eruption involving siliceous magma is the extrusion of magma which is too stiff to flow due to loss of much of its volatile components, and instead swells out to form a dome-shaped mass capping the volcanic pipe (*Plate* 33).

Many of the most violent and catastrophic eruptions are associated with siliceous magmas. The eruptions in 1902 of Mont Pelée in the Caribbean island of Martinique deserves special mention, not

Plate 30   Pillow lavas formed by eruptions in the sea at the foot of Mount Etna, Sicily. These have now been uplifted to be exposed in a sea cliff. *Photograph: J. E. Guest.*

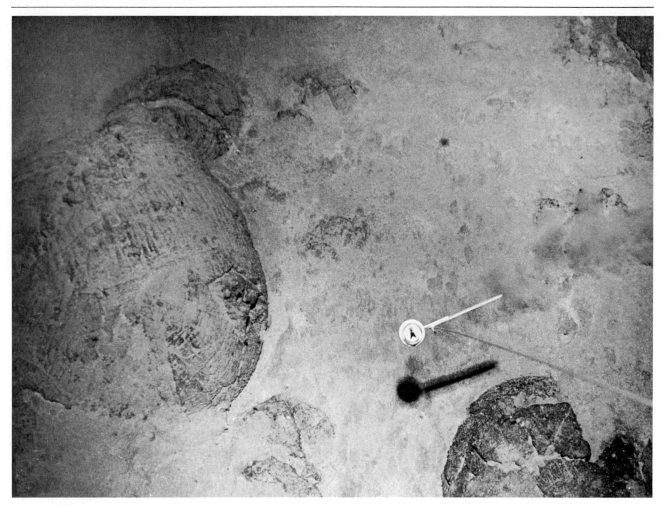

Plate 31 Pillow lavas, partly covered by mud, on the sea-floor at a depth of about 2,800 metres below sea-level off Hawaii. The floating compass is 33 cm long. *Photograph: J. G. Moore, US Geological Survey.*

only because it caused one of the most sudden and complete disasters in history by destroying the town of St Pierre and all but two of its 30,000 inhabitants in a matter of two or three minutes, but also because of the nature of the eruption and its significance in providing an answer to the problem of the origin of ASH FLOWS, or IGNIMBRITES.

As the name implies, ash flows are composed of particles and rock fragments which have accumulated (*Plate* 34) not by falling through the air but by some form of flow. In many respects they have the character of lavas. Commonly the lower part of an ash flow is sufficiently massive and compacted to develop columnar jointing (*Plates* 35, 36). The main components are tiny, irregular-shaped fragments of glass called shards (*Plate* 37), which in the case of the compacted rocks were obviously very hot and still plastic on arrival at their final resting-places, because the shards have become welded together. The upper and generally less compacted part of one of these units might well be taken for an ash-fall deposit except for two things: the fragments are generally neither sorted nor stratified, but instead are stirred into a uniformly heterogeneous mixture; and the top of an ash flow normally has a plane surface extending over a wide area. In fact, the dimensions of some ash flows would do credit to plateau basalts.

Since glassy rocks normally become minutely crystalline with age and features such as shard boundaries become obscured, it is not surprising that ancient formations originating as ash flows used to be regarded as normal siliceous lavas, mostly referred to by the non-genetic name 'rhyolite'.

Now we return to the Mont Pelée eruptions of

Plate 32 An air photograph of an andesitic volcano in the Andes of northern Chile. This volcano has classically symmetrical cone, and its summit is at near 6,000 metres above sea-level. Note the lobate lava flow fronts at the foot of the cone.
Photograph: Instituto de Investigaciones Geologicas, Chile.

The Earth and Its Satellite

Plate 33 (*above*) A volcanic dome in Turkey, produced by the eruption of viscous rhyolitic lava. The lava was too viscous to spread as a flow, and piled up to give a dome over the vent. *Photograph: R. Mason.*

Plate 34 Unconsolidated ash flow tuff (or ignimbrite) rich in blocks of pumice. *Photograph: J. E. Guest.*

Plate 35 An ash flow tuff sheet (or ignimbrite) exposed in a valley side. The zone with well-developed column jointing consists of welded tuff. Southern Peru. *Photograph: J. E. Guest.*

Plate 36 The tops of columnar joints in a welded ash flow tuff (or ignimbrite). *Photograph: J. E. Guest.*

1902 for an explanation. The annihilation of St Pierre was caused by a cloud of very hot gases (up to 800 °C), densely charged with ashes and blocks of lava, which burst out from near the summit of the volcano and rushed down on the city with the combined ferocity of a hurricane and an avalanche. It seems fairly certain that eruptions of this kind, known as *nuées ardentes*, are responsible for the formation of ash flows (*Plate* XIX). The lava particles are kept hot and isolated from contact with one another by the gases, so that though the individual particles may be extremely viscous the net effect is to produce a suspension which has a very low viscosity, able to spread very fast and over large areas.

Craters and Calderas

The feeder of a simple and moderate-sized volcano will normally be in the form of a pipe only a few tens of metres in diameter. This diameter expands rapidly near the surface to give a crater which may be several hundred metres across, formed by a combination of ejection of lava and rock, by piling up of débris round the crater rim and by erosion and landsliding subsequently cutting back the walls. Craters formed in this way are rarely more than 1 km, or at most 2 km, wide. Their shape and appearance naturally depends on the character of the volcanic activity, giving in one instance a deep pit with active lava at the bottom, and in another the crater may be relatively shallow with the floor coated in ash, and with gases seeping through in places, as in the famous example of Solfatara near Naples, where conditions have remained constant since Roman times.

Calderas can be regarded in some ways as outsize craters, with diameters ranging generally from

Plate 37 (*above*) A microscopic view of an ash flow tuff (or ignimbrite) showing the small glass shards of which it is composed. The largest fragments are less than 1 mm across. *Photograph: J. E. Guest.*

Plate 38 (*opposite*) A microscopic view of a lava of the same composition as the ash flow tuff in Plate 37. Here there are no glass shards and the rock has crystallised to a matrix of fine crystals. The large crystal is a plagioclase feldspar. *Photograph: J. E. Guest.*

about 5 to 30 km. However, calderas are not simply products of bigger explosions than craters: they are formed by a combination of factors among which collapse of the floor is generally very important. Evidence for the mechanism comes mainly from the study of areas of ancient volcanicity where structures of the roots of volcanoes are exposed by erosion. Subsidence occurs along cylindrical fractures (conforming roughly to the outlines of the calderas) caused initially by pressures built up in the roof of the magma chamber. Pressures adequate to fracture crust several kilometres thick are most commonly associated with highly siliceous, gas-charged magma. The escape of great volumes of gas, ash and *nuées ardentes* during eruption releases the pressure and may lead to subsidence of the whole of the volcanic superstructure and supporting crust bounded by the caldera ring fault. Sometimes distinctive strata which originally extended across the area of subsidence can be identified both inside and outside the subsided block, showing that the total vertical displacement may be 1,000 metres or more.

Plate XXV
The Hadramaut plateau, southern Arabia. The plateau is dissected by a classical dendritic pattern of youthful valleys. *NASA Gemini 7 photograph.*

Plate XXVI
The coast of the Yucatan peninsula, Mexico, showing a bay with an offshore bar. Note that the detail of the sea-floor can be distinguished. The land area is covered by tropical rain forest. *NASA Gemini 5 photograph.*

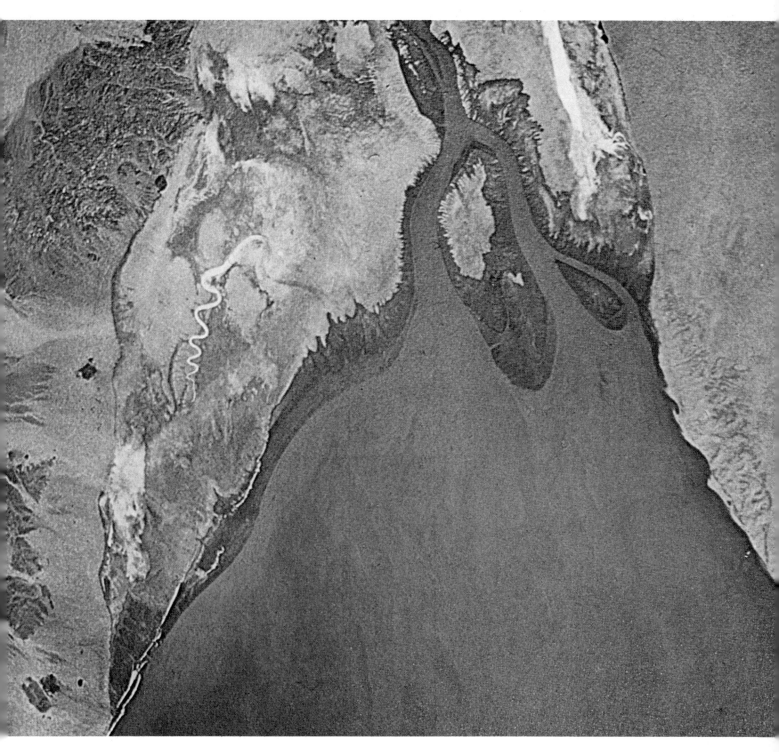

Plate XXVII
The mouth of the Colorado river, USA, where it empties into the Gulf of California. The desert area to the right is the Sonora desert. *NASA Gemini 4 photograph.*

Plate 39 Volcanic bombs scattered over the flanks of a volcano in Iceland.
Photograph: G. P. L. Walker.

Plate 40 An open volcano fissure in northern Iceland. *Photograph: G. P. L. Walker.*

Plate XXVIII (*opposite*)
The southern tip of Florida, USA. The shallow-water reefs to the right are clearly seen. *NASA Gemini 4 photograph.*

Chapter Six

Surface Processes on Earth

A. J. SMITH

The same external forces act to change the face of the Earth as act to change the face of the Moon. These forces are gravity and solar energy. Because of the temperature and pressure conditions which exist, and have existed from the earliest times of Earth's history, this planet has, in contrast to the Moon, an ATMOSPHERE and a HYDROSPHERE. These two, which are closely related, act together to modify the effects of solar energy and give rise to a radically different set of weathering and depositional conditions from those on the Moon. The principal rôle on Earth is played by water.

Heat from the Sun evaporates sea-water and, using the winds in the atmosphere as its agent, also provides the energy that causes the circulation of water vapour by winds; some of the water vapour is precipitated as dew, fog, rain and snow on the land areas of the planet. These may in turn be evaporated again or flow as streams, rivers or glaciers back to the sea. This circulation of water is significant in moulding the landscape of the planet. Water helps in *weathering*—the breakdown—of rocks on and near the surface; water lubricates the downhill movement of the weathered material. Water as rivers removes weathered material, and carves deep valleys in the landscape; water in its solid form as ice also acts as an eroding agent. Even in arid regions water still plays an important rôle. It also dissolves rocks, some of which are more easily dissolved than others, and carries the dissolved salts to the sea in solution.

The sediment load carried by water is eventually deposited in the sea, the elements in solution combine and are directly precipitated or are removed from sea-water by living organisms to be part of the accumulation of sediment on the sea-floor. These sediments, by compaction and cementation, become lithified to give the sedimentary rocks. Earth movements and changes in sea level bring these rocks into contact with the atmosphere, and weathering and erosion commence again. This is the weathering cycle.

The weathering process and subsequent erosion give rise to landforms on the planet which can be classified according to the principal erosive agent. The landforms change, for the most part, slowly, though catastrophic changes occur from time to time.

Weathering

Weathering is the process of change which affects rocks at and just below the Earth's surface and leads to their decomposition and disintegration

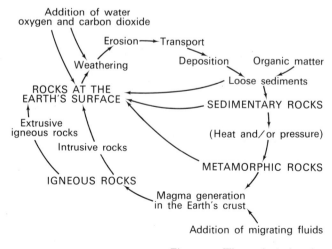

Figure 19   The geological cycle.

Surface Processes on Earth

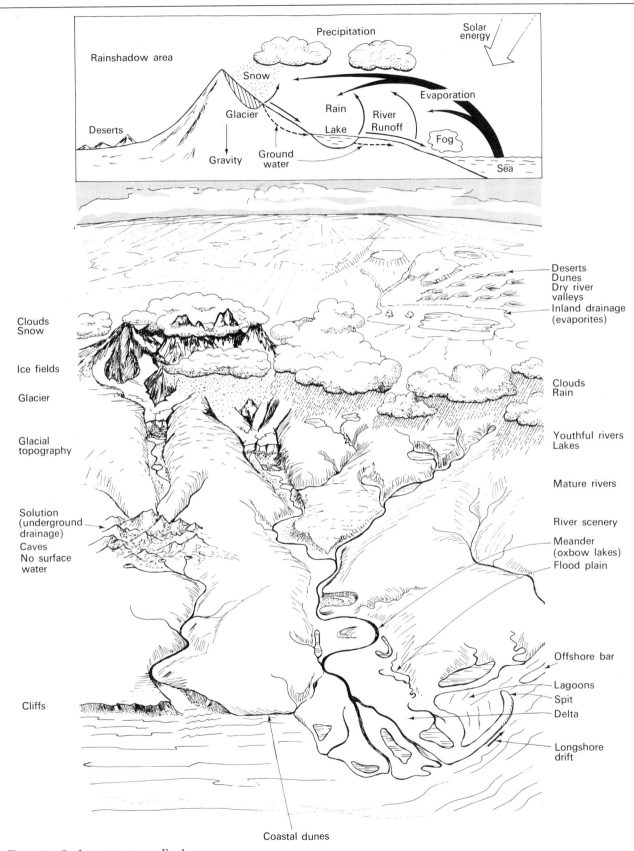

Figure 20 Surface processes on Earth.

67

(*Plate* 41). The changes take place at normal temperatures and pressure, and water plays an important rôle.

Two general types of weathering are recognised: *physical* (mechanical) weathering, and *chemical* weathering, leading to disintegration and decomposition respectively—though usually both types may be occurring at one and the same time, e.g. changes in volume during chemical weathering hasten physical disintegration.

Physical weathering is common at high altitudes and high latitudes where water may freeze and thaw in a daily or seasonal cycle. The volume change which accompanies the freezing of water prises the rocks apart. Plants in all climates can exert a physical effect, their roots prising rocks apart. However, plants also exert a chemical effect by holding water and by creating humic acid. Humic acid and dilute carbonic acid, rainwater with carbon dioxide taken from the atmosphere, increase the ability of water to dissolve rocks. Some rocks such as rock salt (table salt) dissolve readily; others such as limestone are prone to solution.

No rock-forming mineral is inert against the effects of water, oxygen and carbon dioxide of the atmosphere. Of the common rock-forming silicate minerals OLIVINE is the least stable, then CALCIUM PLAGIOCLASE, PYROXENE, HORNBLENDE, SODIC PLAGIOCLASE, BIOTITE MICA, ORTHOCLASE, MUSCOVITE MICA and QUARTZ, which is the most stable. Even quartz is slightly soluble in fresh water, more so in saline water.

Chemical weathering is faster in humid areas with high temperatures; the products of weathering in such areas are more complex, important resources such as *bauxite* (a source of aluminium) being produced in this way. In desert areas water, perhaps surprisingly, plays an important role in weathering.

The effect of weathering is to produce a zone of altered rock with some of the original rock in solution in ground-water enclosed in the pores. The thickness of this zone depends on rates of weathering and the speed of removal of the products of weathering from the surface.

The products of weathering, with the addition of organic matter, give rise to the soil cover of the planet upon which so much of man's livelihood depends.

Erosion

Weathering is a static process—it breaks down rocks and prepares them for erosion, a dynamic process.

On Earth, as on the Moon, gravity alone can move the mantle of weathered material. The process is often referred to as MASS WASTAGE and, on Earth, the lubricating effect of water speeds the process.

On really steep faces, rock fragments may just fall

Plate 41 Gravestones, Newcastle upon Tyne, England. These stones, with lettering carved about 150 years ago, are beginning to show signs of weathering, particularly near ground-level. A useful method of checking on the durability of various rock types is to note the dates and weathering on gravestones. *Photograph: J. E. Robinson.*

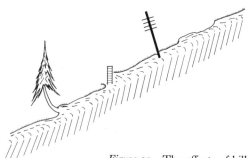

Figure 21 The effects of hill creep.

away to form a scree below (*Plate* 42). Abrupt rock falls and landslides, which may be triggered by earthquakes, occur in high mountain areas each year. Sometimes these slides will be mixtures of rock, water and ice and can have catastrophic results. On lower slopes, masses of rock and soil may slide or flow (*Plate* 43) even on grass-covered slopes in temperate areas; that creep of the surface takes place is evidenced by leaning posts, curved tree-trunks and the bending of rock planes (*Figure* 21).

Plate 42 Screes at Bow Lake, Banff National Park, Canada. The screes were created by frost shattering after the retreat of the Crowfoot Glacier, which can be seen in the background. The lake was formed by the overdeepening of the valley by glacial erosion. *Photograph: P. G. Llewellyn.*

Plate 43 Mudflow, Isle of Wight, England. The clay, in this case London Clay, has moved forward due to wetting by heavy rains. *Photograph: J. E. Robinson.*

Plate 44 Earth Pillars, Gerome, Anatolia, Turkey. The hard cap rock protects the more easily eroded underlying material from complete denudation. Such features usually occur where erosion is by rain, the hard cap rock acting as an umbrella to protect the softer material. *Photograph: A. J. Smith.*

Flowage of weathered material is particularly effective in waterlogged areas subject to freezing and thawing; this process of movement is called *solifluction*, and is a common phenomenon in arctic regions.

All these mass movements tend to bring loose rock debris within reach of streams or glaciers which continue the transport to lower ground and, eventually, to the sea. Streams, by eroding the foot of slopes, make them unstable and thus speed the process of mass wastage on slopes.

Geological Work of Rivers

Rivers are the most important agents that transport the products of erosion; they drain vast areas and, even in deserts where flow is infrequent, they can carry more material than the wind.

As raindrops hit the ground they aid the movement of weathered particles down-slope. The rainwater, as it seeks lower ground, carries particles with it. Vegetation cover will control how much débris is moved in this way, but the rainwater, as it concentrates in depressions will, on any slope, soon cause channelling (*Plate* 46). Permanent streams develop when run-off and underground water combine. As streams join together, forming rivers, their erosive power is increased.

If the bedrock is soluble, solution will occur. Hard, resistant bedrock is eroded by the abrasion of particles moved by water; a common type of abrasion results in cylindrical holes cut in the bedrock by stones spun in the eddying currents. These holes are called *potholes* (*Plate* 50). Softer bedrock and loose material can be moved by the pressure and shearing effect of the moving water.

The visible load of any river is carried in suspension, by bouncing, or by traction along the stream bed. How much a river can carry depends upon the volume and velocity of water. Rivers in flood are capable of moving great loads including large boulders. A fall in volume or velocity leads to deposition.

In mountainous areas rivers expend much of their energy deepening their valleys, leading to characteristic V-shaped valleys (*Plate* 47). The profile, the longitudinal section, of the river at this stage is irregular, reflecting variations in rock hardness and the imbalance between the river's capacity to erode and the load to be carried. Waterfalls (*Plates* 48, 49), lakes and boulder-strewn valleys are typical.

Downstream, where the river becomes larger due to the confluence of tributaries, more energy is used widening the valley. The size of particles has been reduced by erosion and attrition, but the load can be considerable. The longitudinal profile will be smoother and less steep and the river will begin to develop a flood plain (*Plate* VII). Still farther downstream, as the river approaches base-level, usually sea-level, the gradient is much less and the river now carries only fine particles. At this stage the river meanders in its flood plain and its channel may even be raised above the valley floor with LEVÉES banking the river. In these lower reaches the valley is very wide. The action of running water is to reduce even mountainous areas to a PENEPLAIN. Rivers rarely complete the erosion of a land area to a plain before earth movements lead to an uplift of the

Plate 45 Inselbergs, Chad Republic. The photograph shows the isolated hills in the late stage of erosion under arid conditions. Inselbergs are typically steep sided because a great deal of erosion takes place at the foot of the hill. Spread over the foreground are the products of erosion. *Photograph: R. J. G. Savage.*

land and a rejuvenation of the river system. But for such earth movements rivers could remove all features from the continents in, it is estimated, 44 million years. The Mississippi (*Plate* 73) illustrates the tremendous load rivers can move—some 500 million tons of sediment and 136 million tons in solution are carried into the Gulf of Mexico each year.

Underground Water

Some rainwater soon returns to the atmosphere by evaporation and some runs into streams, but a considerable proportion, depending upon climate, vegetation and the nature of the land surface, percolates into the weathered zone and the underlying rocks to become ground-water. This water, when it can percolate no farther, fills all the spaces in the rock and weathered material and may emerge as a spring. Many parts of the world have great amounts of water held below the surface. This water will dissolve soluble rocks, aiding the weathering process. Some rocks, such as limestone, are prone to such solution by surface and underground water, and a characteristic landscape develops in such areas—dry valleys, cavern systems, sink holes and associated phenomena are developed. This landscape is called *karst* topography after the name given to such a region in Yugoslavia.

Geological Work of Ice

The polar regions and high mountains of Earth have ice-fields. Precipitation is in the form of snow, and where snowfall exceeds the loss by melting the

Plate 46 Incipient gullies in loess, Otago, New Zealand. These are the first stage in the development of river valleys. *Photograph: D. Hamilton.*

Plate 47 (*below*) River gravels, Haast river, New Zealand. The gravels have been swept down from the mountains at time of flood. The river at normal times is incapable of transporting such a load, hence these large deposits. *Photograph: D. Hamilton.*

Plate 48 Waterfalls, river Mellte, South Wales, showing the erosive force of a small stream when in spate. The falls are caused by the rejuvenation of the river system. *Photograph: A. J. Smith.*

snow is compressed into ice which can move under its own weight to lower ground. The ice at the polar regions may exceed 3 km in thickness. In high mountains the ice is less thick but moves more rapidly to lower ground forming valley glaciers. Glaciers modify the landscape by erosion and deposition.

Much of the scenery of the high mountains of the world is due to ice action; pointed peaks (*Plates* 57, 58), sharp ridges and chairshaped valley-heads are typical features. When ice has retreated we see a much modified landscape. Rock surfaces are rubbed smooth by moving ice (*Plate* 60) and rock-embedded ice, U-shaped valleys with hanging tributaries develop (*Plate* 59), and the interlocking spurs, so typical of stream-cut valleys, are removed. Local over-deepening creates lake basins (*Plate* III).

Deposition takes place where the arrival of ice is balanced by melting; at such a place the load of rock and ground fragments is deposited as a moraine, and in addition the meltwater from the ice leads to great outwash plains of sediment.

So characteristic are many of the features of glaciation that they stand as irrefutable evidence of greater past glaciations. Commencing about a million years ago the ice-cap of the Arctic spread out over northern North America, north Europe and Asia, while the mountain glaciers of the Rockies, Alps, Himalayas and other mountain chains were much larger and reached lower altitudes. Such changes led to a lowering of the sea level and ice covered many sea areas, including the Baltic and the North Sea.

There is evidence of earlier ice ages in the geological past—parts of southern Africa, India and Australia were affected by ice 275 million years ago (*Plate* 60). Such evidence is an important indication of past climates and the wandering of continents and poles.

Geological Work of the Sea

Where the sea meets the land its restless energy is at work modifying the shoreline. The sea can erode the land with great effect (*Plate* 52) and large areas of land have disappeared in historical times. Some of the products of erosion are swept into deeper water, but a considerable amount is moved along

Plate 49 Iguacu falls, Iguacu river, on the boundary between Argentina and Brazil. The falls occur where the river leaves the hard Parana Basalts. Total height of the falls about 80 metres. Note the gorge caused by the retreat of the falls. *Photograph: P. G. Llewellyn.*

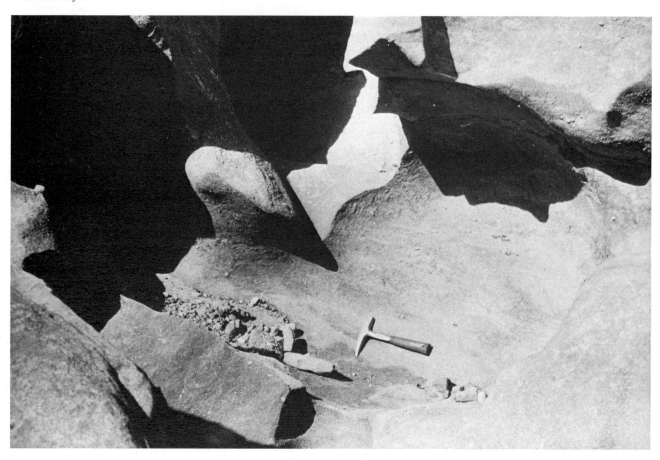

Plate 50 Potholes, Bheemeram, central India. These cauldron-like features are formed by boulders and pebbles being swirled around by flood waters when the stream bed is filled in the rainy season. *Photograph: A. J. Smith.*

Plate 51 Alluvial fan, Haiti. The cone of sediment has been built up by sediments carried out from the mountainous region at a faster rate than the sea can erode the sediment. *Photograph: A. J. Smith.*

Plate 52 (*below*) Sea stack, Sandfly Bay, Otago, New Zealand. The photograph dramatically shows the force of a breaking wave. In the background a sea stack of hard rock. *Photograph: D. Hamilton.*

the shoreline by currents to add it to the coast in another place.

How effective the sea is depends upon the hardness of the rocks, the force of the waves, the degree of exposure, the depth of water near the shore, the effectiveness of longshore currents and the proximity of rivers bringing heavy sediment loads into the sea.

Like streams and rivers, waves need tools to be effective. The first blocks are prised off by wave action alone, then the waves hurl the blocks at the foot of the cliff, eventually undercutting it, and causing it to collapse (*Plates* 53, 54). The collapsed débris is removed by wave action and currents and a wave-cut platform is created. There is a limit to the size of this platform when sea level remains static, for the platform absorbs energy from the waves by friction. However, should sea level be rising—and sea level has risen and fallen several times on a world scale in the past million years due to changes in the volume of ice in polar regions—then extensive wave-cut platforms can be created. Any irregularities in the platform will be filled by the products of erosion. The flat surface of the continental shelves is believed to be the combined effect of erosion and deposition. There are also level platforms to be seen on land which were cut by the sea when sea level was higher in relation to the land than it is now.

The energy of waves decreases rapidly with depth and, farther offshore, except where there are strong tidal currents, only fine sediment is kept in motion. Much of this sediment is carried off the edge of the continental shelves into deeper water.

Erosion in Arid Regions

Regions receiving less than 25 cm of precipitation each year are called deserts. Here, for much of the year, wind is the main erosive force; it carries the products of weathering away from some areas leaving bare rock. The wind, armed with sand grains, erodes rock surfaces and boulders by sand-blasting.

Plate 53   A wave-cut arch in limestone, Malta. *Photograph: A. J. Smith.*

Much of this erosion is just above ground level, leading to the formation of steep cliffs and faceted pebbles. The grains can be moved hundreds of kilometres in swirling dust-storms and large parts of deserts are covered by moving sand dunes (*Plate VI*). Some of these dunes may be more than 300 metres high: the Tifernine Dunes of eastern Algeria in the Sahara reach 500 metres. The finest particles may be carried out of the desert region to be deposited as loess or brickearth.

Wind is an important erosive agent in areas bordering glaciated regions where there is little free water; during the Pleistocene glaciations wind-borne sediments were distributed over wide areas of North America and central Europe. To a lesser extent wind erosion and transport is important in areas bordering large beaches where the prevailing winds are on-shore.

In spite of the effectiveness of wind the infrequent, but usually extremely heavy, rain-storms in deserts can lead to severe erosion. The lack of vegetation cover permits rapid erosion and the run-off carries vast amounts of debris in the short-lived sediment-laden streams which flow down the normally dry wadis. The streams are often choked with sediment and deposition occurs in large alluvial fans as soon as the river descends to a plain.

The Net Result of Erosion on Earth

Earth, with its atmosphere and hydrosphere, has created on its surface distinctive landforms. Each agent is capable of creating characteristic landforms, and changing climatic controls can lead to differing landforms being superimposed. Although the total effect is to reduce the land surface to flat plains, earth movements and long-term changes in sea level are constantly changing the elevation of the land areas so that erosion keeps fashioning new landforms, but never to completion.

Sedimentation

The significant, indeed the principal, rôle in shaping the land surface of the Earth is played by water—even in deserts water is an important agent of erosion and transport. Water erodes the surface of the planet and transports the products of weathering and erosion to the sea. In the sea, as is described below, vast volumes of sediment accumulate and by compaction and cementation the sediments are

Plate 54   A natural cave formed by sea erosion along a zone of weakness in the rock. Note the wave-cut cliffs in the background. *Photograph: A. J. Smith.*

made into hard rock. By earth movements these rocks are raised above sea level and the processes of weathering and erosion commence work on them again.

Three major environments of deposition can be recognised: the *continental environment*, where sediments are laid down on land, though most of these sediments may later be re-eroded and transported to the sea; the *marine environment*, where sediments are deposited on the sea-floor; and the *mixed continental-marine environment*, that is, where the continents border the sea.

Continental Deposits

A wide variety of sediments are deposited on the continents for variable lengths of time; indeed, by tectonic accident some continental deposits are preserved and can be found in the geological record.

The sediments deposited on the continents are usually related to the method of transport. *Fluviatile sediments*—the deposits of rivers—are extremely varied; in mountainous regions they range from huge boulders to fine sand, such deposits being local and usually temporary, often moved by the next flood. Where rivers leave the mountains great fans of sediment frequently build up (*Plate* 51), and stream courses may be choked by sediment. Farther downstream fluviatile sediments tend to be less coarse-grained, better sorted and the grains more rounded. Indeed any river section when traced downstream shows a progressive reduction in size of the largest particles and an improvement in the degree of rounding of the grains.

In their lower reaches rivers build extensive flood plains. The river moving slowly often reworks its own deposits. As the river extends its course out to sea with a delta it has to raise its riverbed in the lower sections of the valley to maintain the low seaward gradient. Some rivers, by building up natural levées—banks created by the deposition of river sediment when a river leaves its course in flood—are raised above their flood plains. When such a river floods, flood waters spread out over the flood plain and the sediment load of the river is deposited over a wide area.

A lowering of the river's bed, caused by normal erosion in the upper reaches or by the uplift of the land or a lowering of sea level affecting the lower reaches, can lead to the flood-plain deposits being eroded and remnants left as terraces above the valley floor.

When glaciers and ice-sheets reach the zone of wastage they begin to drop their load of debris. When, due to the amelioration of the climate, the ice retreats, glacial deposits are left stranded on the landscape. Glacial deposits and the associated fluvioglacial deposits, i.e. the deposits of sediment carried by meltwaters, are grouped together under the general term *drift*, and such deposits cover a great part of North America and Europe as a consequence of the Pleistocene glaciation. The commonest type of drift is *boulder clay* or glacial *till*, which consists of unsorted and unstratified sediments ranging in size from fine clay to boulders of immense size. Most of the stones are irregular, a few are flattened and grooved. Many of the boulders and stones may differ in character and composition from the rocks on which they rest. These boulders are called *erratics* and in some cases can be traced over scores of kilometres to their source.

Glacial deposits tend to be thickest where the glacier advances into the zone of wastage. The resultant accumulations of drift from glacial ice have distinct topographic forms—collectively called ground moraine and end moraine. Ground moraine occurs as a thin sheet over the landscape filling in pre-glacial landscape and covering the solid geology. End moraines mark the end and stages in the retreat of the glaciers. Beyond the terminal moraines of ice-sheets and glaciers are large areas of fluvioglacial deposits spread out by meltwater streams. The sediments are poorly sorted but are stratified.

Aeolian wind-borne deposits are associated with warm and cold deserts. Sand dunes are a characteristic deposit and many large deserts have sand seas. The finest particles can be carried by the wind vast distances beyond the deserts. These fine sediments are called loess and extensive deposits were created during the Ice Age, for at that time there was little free water at the surface and fine particles were easily picked up and transported by the wind. Smaller developments of wind-borne deposits occur near large beaches exposed to prevailing on-shore winds—the dunes of the Landes region in south-west France are an example of this.

Lakes and swamps are especially numerous in the wide continental areas that were subject to glaciation in Pleistocene times. Lakes fill with sediment in a relatively short time, and many of them received large amounts of sediment from melting glaciers. These deposits, when they show seasonal banding, are called *varved* sediments. In temperate and humid conditions, organic sedimentation plays an important rôle in filling a lake—indeed many lakes in temperate areas receive more organic sediment than clastic sediment. As rivers continue to erode they will eventually lower the exit of a lake and even cut through the lake sediments.

In arid regions some rivers may not reach the sea and areas of inland drainage develop. The intense

Plate 55 Ripple mark, Kistampet Sandstone, central India. These features were formed in shallow water by wave action at the time this rock was deposited about 800 million years ago. The sand was buried and lithified, only to be exposed again in the present day. *Photograph: A. J. Smith.*

Plate 56 Flute marks, Aberystwyth Grit Series, Cardiganshire, Wales. These marks were cut into muddy sediments by a turbidity current, filled with sediment and so preserved. The marks are on the underside of the sandstone bed; the block of rock in this illustration is upside down. *Photograph: A. J. Smith.*

Plate 57 Gorner Glacier and Matterhorn, Alps. This picture shows a variety of interesting glacial features—the pyramidal peak of the Matterhorn, the high snow-fields, the glacier with its crevasses and lateral and medial moraines. The foreground is typical of recently glaciated country—a great deal of bare rock with scattered boulders. Photograph taken about 1870. *A. J. Smith collection.*

Plate 58 Gorner Glacier, with Matterhorn in distance, Alps. Photograph shows large boulder which has been transported by ice. Note how the boulder, by giving shade, is protecting ice from melting. (Compare with the photograph of Earth Pillars, Plate 44.) Photograph taken about 1870. *A. J. Smith collection.*

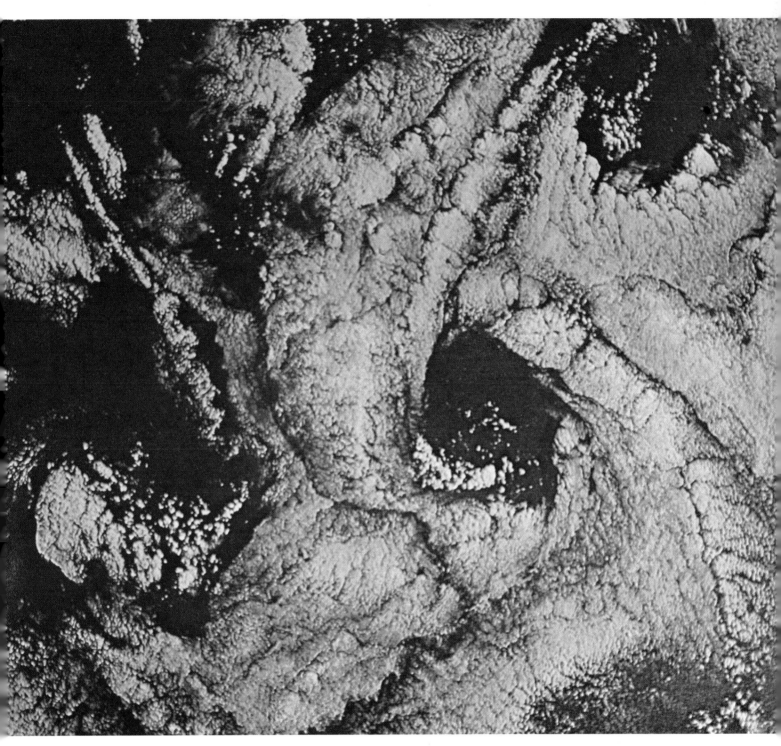

Plate XXIX
A spiral formation of strato-cumulus clouds in the Atlantic south of Tenerife, one of the Canary Islands. Winds blowing from the north are deflected by Tenerife to give large eddies such as this one. *NASA Gemini 6 photograph.*

Plate XXX
Cloud formations over China in the region of Chengtu and Changsha; altitude 150 km; horizon distance 1500 km. *NASA Mercury 9 photograph.*

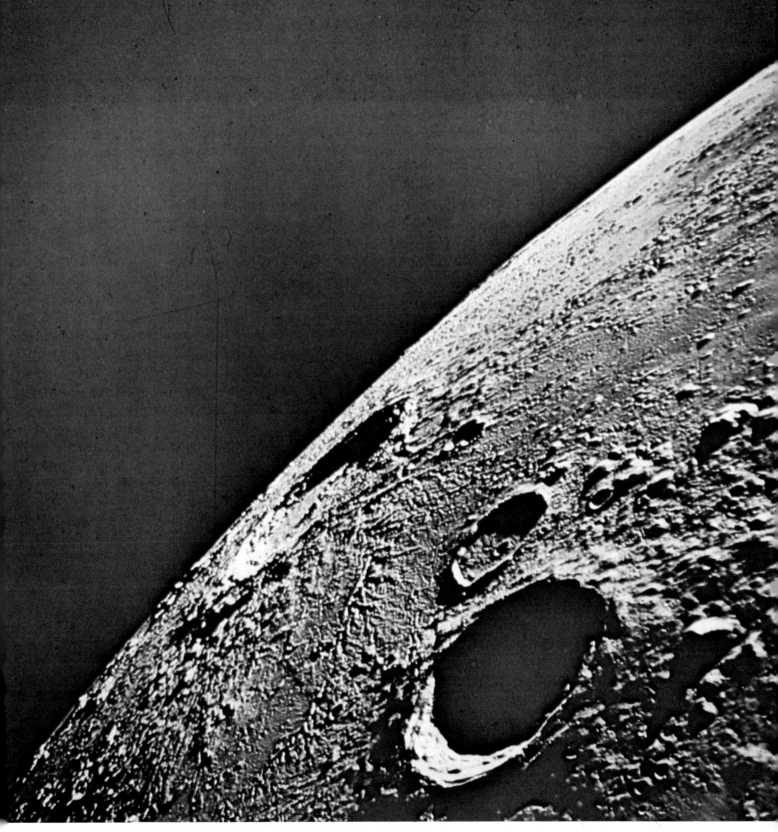

Plate XXXI
Oblique view of lunar surface towards Copernicus with Reinhold in foreground; altitude 100 km. *NASA Apollo 12 photograph.*

Plate XXXII
Comparison of Earth and Moon photographed from space vehicles; the Earth was photographed during the Apollo 10 flight of March 1969; the Moon was photographed during the Apollo 11 flight of July 1969; the disks are reproduced to show correct relative diameters. *NASA photograph.*

Plate 59 Glaciated valley, near Jungfrau, Switzerland. Typical U-shaped valley eroded by a now retreated glacier. Note the steep sides, the hanging valley with waterfall and the alluvial cones. *Photograph: E. J. W. Jones.*

evaporation leads to the chemical precipitation of minerals and glistening deposits of rock salt, gypsum and other salts are formed—the *salinas* of the western United States.

Deposits found in caves are often scientifically interesting because of the well-preserved fossil remains, including those of early Man, which they sometimes contain. They consist of sediments brought in by floods and those formed by the precipitation of minerals from solution, the characteristic stalagmites and stalactites.

Mixed Continental-marine Deposits

These include the deposits of deltas, estuaries and beaches.

Deltas of great rivers bear vivid testimony of the transporting power of rivers. As all large deltas slope gently outwards under water, their real area is several times greater than that above sea level. They often consist of very thick piles of sediment—

Surface Processes on Earth

the Mississippi delta, the best studied of all the great deltas, has a maximum thickness of more than 8 km, a clear indication that in this case there has been considerable crustal depression.

The rate at which deltas build out into the sea can be measured by direct observation and by comparing old charts. The rate of seaward growth may be as much as several kilometres in each century. Naturally, the rate of development will depend upon the amount of sediment brought down by the river and the rate at which the sea can move it.

Deltaic deposits have a complex variety—river muds and sands, lake and marsh sediments deposited between the distributaries and marine deposits. The complexity is due to the often abrupt changes in course of the distributaries and the fact that changes in sea level can cause the sea to flood over a delta at times. Most deltas consist of fine sediments, but some deltas of large, fast-flowing rivers may contain quite coarse sediments, as is the case in the Rhône delta.

Sediment from some deltas and from rivers which do not have deltas may be moved laterally by the sea, creating spits and bars. Behind these develop lagoons which may, in humid and temperate climates, contain fresh or brackish water and have an extensive growth of vegetation in and around them. Such conditions on a large scale prevailed in the northern hemisphere in Carboniferous times and gave rise to thick plant deposits now mined as *coal*.

Not all large rivers have deltas; many have estuaries (*Plate* XXVII). These estuaries have a distinctive range of sediments, mainly muds, but characterised by fresh, brackish and marine faunas, depending upon the place in the estuary.

A significant area of mixed continental and marine deposition is the low-lying areas which are subject to flooding by tidal waters or by exceptionally high tides. In temperate regions tidal flats are developed behind off-shore bars. In arid areas evaporite deposits usually occur. A particular case is those areas subject to occasional flooding such as the *sabkhas* of the Persian Gulf. Here a wide variety of evaporite sediments are being formed.

Last in this category are beaches—the line between land and sea where the restless energy of the sea is so apparent in the waves which sort and round the pebbles and sand grains so effectively and where the movement of the tides can be clearly seen.

The presence of raised and submerged beaches is

clear evidence of changes in sea level. Sea level is not a fixed datum in any geological sense—because of earth movements and because of changes in the volume of ice in the polar ice-caps, great variations can take place in sea level. There is ample evidence in the succession of strata in the geological column of episodes of *transgression*, when sea level rose relative to the land, and of *regression*, when sea level fell. The deposits related to a transgression show an upward change from terrestrial or mixed continental-marine deposits through beach deposits to marine deposits laid down in progressively deeper water, younger deposits overlapping older deposits.

Related to the advance and retreat of the Pleistocene ice-sheets there have been several quite rapid changes in sea level in the past million years. Indeed, the last significant rise in sea level about 6,000 years ago has left its mark not only in its deposits but in the folklore of many ancient peoples.

Major transgressions and regressions in the past can be associated with earth movements, either broad regional warpings of the Earth's crust or more dramatic folding and faulting caused by mountain-building forces.

Marine Deposits

The deposits of seas and oceans can best be described by making two subdivisions. The first is based on the type of sediment, that is into *clastic* (created by the erosion of other rocks) or *biochemical-chemical* (derived from sea-water by sea creatures or by direct precipitation from solution). The former includes pebbles, sands, clays and shell sands; the latter coral reefs, organic oozes, salt deposits and manganese nodules. The second method of subdivision depends upon the depth of sedimentation: on the one hand sediments deposited in shallow water, less than 200 metres and, on the other, sediments deposited in deep water. The former includes the deposits of the continental shelves and where light can penetrate to the sea-floor, the latter the deposits of the continental slopes and ocean floor.

Clastic deposits predominate near land; the coarsest sediments are often found bordering the shore while, generally speaking, finer sediments are found farther from the land. The rate of sediment supply, the size of particles in that supply and marine currents will tend to modify such a pattern.

Organically formed sediments are common in many shallow-water areas where terrigenous (land-derived) sediments are less abundant. The type and quantity of organic deposits will depend upon the class of creatures which happen to predominate, the amount of nutrients, the amount of sunlight which penetrates the sea and the sea temperature. In shallow water the marine creatures which contribute on the greatest scale are collectively known as *benthos*—bottom-dwellers, including all forms which live on the sea-floor whether attached to it or not. In the warm, clear waters of the tropics, with upwelling waters from the oceans providing abundant nutrients, we get the development of coral reefs, ranging from fringing reefs through barrier reefs to atolls.

Even where terrigenous clastic deposits are abundant, benthonic creatures and their fragmentary remains add significantly to the total amount of sediment deposited. In many ancient sedimentary deposits these organic remains, the fossils and fossil fragments, are an obvious constituent of the rock.

Marine deposits predominate in the succession of strata and many of these deposits were laid down in shallow water. Such successions could have been formed only when the shallow sea-floor was sinking at about the same rate as the deposition of sediments (*Figure* 22).

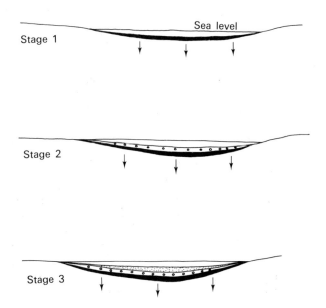

Figure 22 These figures illustrate how considerable thicknesses of shallow-water sediments can accumulate. While the depth of water remains the same, the sea-floor sinks, thereby creating room for more sediments to accumulate.

Surface Processes on Earth

Plate 60 Striated pavement, Irai, central India. This rock surface was sculptured by ice nearly 300 million years ago. It is identical with similar pavements created in the Pleistocene glaciation of about one million years ago. It is one of many lines of evidence used by geologists to elucidate climatic regimes of the geological past. *Photograph: A. J. Smith.*

Plate 61 Folded strata, Jutland, Denmark. These rocks were folded by ice, not by tectonic forces. It illustrates the force exerted by moving ice. *Photograph: A. J. Smith.*

Plate 62 Glacial outwash gravels and sands, Cheshire, England. *Photograph: J. E. Robinson.*

Plate 63 Bedded rocks and wave-cut platform, Aberystwyth, Wales. The picture shows uniformly bedded sedimentary rocks. Note how the sea has planed off these rocks at the foot of the cliff to create a wave-cut platform. *Photograph: A. J. Smith.*

Today only finer sediments are being deposited at the outer edge of the continental shelves, though at times of much lower sea level coarser sediments may also have accumulated there. At times sediments from the outer edge of the shelves slide down the continental slope to become huge slump deposits or, by mixing with sea-water, may become TURBIDITY CURRENTS (*Plate* 56) which sweep down the slopes and across the abyssal plains of the ocean floor adjacent to the continents. Fine terrigenous sediments carried out by ocean currents are also deposited on these plains.

In many parts of the deep oceans the principal sediments are in the form of calcareous and siliceous organic oozes. These deposits are composed of the remains of tiny *plankton*—organisms which live in the surface waters of the oceans, being carried by the ocean currents. These single-celled marine plants (diatoms and coccolithophores) and animals (Foraminifera and Radiolaria) and a variety of higher forms of life have no means of self-locomotion. On death their remains fall to the ocean floor and give rise to the deep-sea oozes. The composition and distribution of these oozes depend upon the temperature of the surface waters and the depth of the underlying ocean floor. The latter is important for with increasing depth many of the remains, particularly the calcareous varieties, pass into solution once again.

In regions free of oozes, only red clay is deposited. This is the product of the finest terrigenous material carried by ocean currents or even by the planetary winds, supplemented by extraterrestrial debris and the most resistant parts of *nekton* (free-swimming creatures), such as the teeth of sharks. In many parts of the red clay there are abundant manganese nodules, the origin of which is still the subject of much scientific discussion.

The Oceans

No account of the form of the Earth would be complete without some description of the submarine morphology of the planet. More than seven-tenths of the planet is covered by seas and oceans, but the rate of scientific investigation of the sea-floor since World War II has been such that a great deal is now known of the Earth beneath the sea.

About 7·5% of the ocean area covers the continents giving rise to the *continental shelves*. Here the water is relatively shallow. The shelf descends slowly oceanwards at a slope of generally less than 1° until there is an abrupt change in slope which marks the transition to the continental slope. This point of transition is called the *shelf break* and it occurs at different depths in different parts of the world—about 50 metres off Florida and about 200 metres off western Europe, while in other areas the break may occur at depths greater than 400 metres. The surface of the continental shelves tends to be nearly flat, being smoothed by a combination of marine erosion and deposition.

Beyond the shelf break are the *continental slopes* which extend down to depths of 1,500 to 3,500 metres. The slope averages about 4° to 7° and is frequently cut by great steep-sided submarine canyons. The origin of the latter is still the topic of much discussion; erosion by turbidity currents at a time of lower sea level is one possible explanation.

Often at the foot of the slope there is a broad zone of sediment accumulation which is called the *continental rise*. Its surface is less steep than the slope and it passes out into the *abyssal plains*. Some continents are bordered by a *marginal plateau*, for example the Blake Plateau off Florida at a depth of between 600 and 1,000 metres. Other continents have no shelves and are bordered by deep trenches, such as the marginal trench off Chile which reaches a depth of more than 8,000 metres.

Away from the continents the ocean floor is by no means flat—there are *island arcs*, often associated with *island arc trenches*—with depths including the greatest deeps on the planet of about 11,000 metres in the Marianas trench in the Pacific. In addition to these features there is a world-wide system of *oceanic ridges* which rise 2,000 to 3,000 metres above the ocean floor. The highest points in the ridges are near the centre of the ridge though the main crest is usually split by a central rift valley. Away from the crest the peaks are less high, passing laterally into the abyssal plains. The central ridge of the Atlantic Ocean is 1,600 to 1,900 km wide.

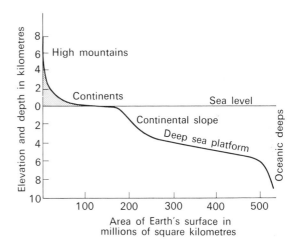

Figure 23 The hypsographic curve showing the area of the Earth's surface above any given level.

Elsewhere on the ocean floor are broad *ocean rises*—the Bermuda Rise of the Atlantic and the Lord Howe Rise of the Pacific are two examples. These rises are distinct from *sea-mounts*, isolated or comparatively isolated elevations rising approximately 1,000 metres or more from the ocean floor, and *guyots*, sea-mounts with flat tops generally deeper than 200 metres from the ocean surface. These features are widespread and are often associated with oceanic rises. A volcanic origin is widely accepted, with the possibility that the flat

Plate 64 Two views of Meteor crater, Arizona. (A) is an oblique air view of the crater; and (B) is a vertical air view. The crater is about 1,200 metres across.

Surface Processes on Earth

top of the guyots was made by erosion at sea level before submergence.

The remainder of the ocean floor is nearly flat with abyssal plains and scattered abyssal hills.

Finally a distinct contribution to the morphology of the oceans comes from coral reefs (*Plate* XXVIII) ranging from fringing and patch reefs to barrier reefs—such as the Great Barrier Reef of Australia— to atolls. Charles Darwin suggested an explanation for the last which envisaged a progressive development from fringing, through barrier to atoll reefs as the land, usually a volcanic island, was submerged by the sea.

This account of the surface features of the Earth, both above and below sea level, has been concerned with natural processes which have continued throughout geological time. However, the surface of the planet is now increasingly being modified by man's activities. In developing his agriculture he has removed surface cover and increased rates of erosion and by his engineering skills he has carved great holes, dammed rivers and covered large areas with a combination of concrete, bricks and tarmacadam; he has pushed back the sea and polluted the atmosphere and oceans. In this way he is playing a geological role which further distinguishes the Earth from its satellite—and presumably from all other planets in our solar system.

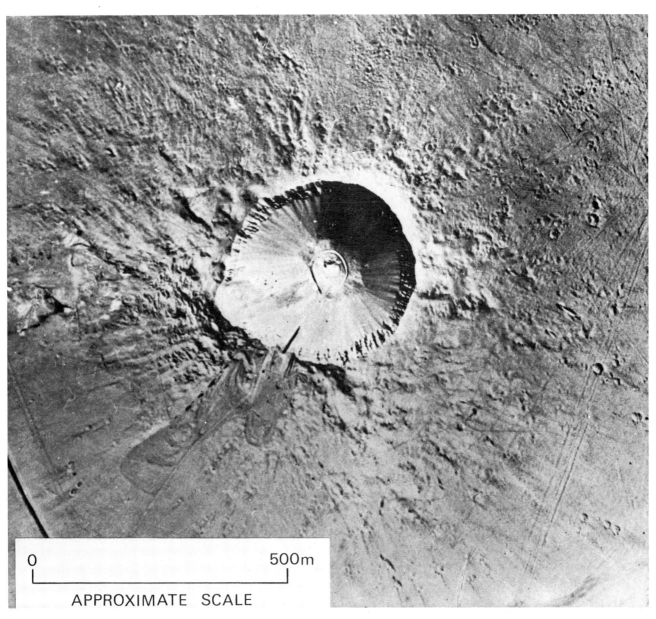

Plate 65 An air view of the Sedan Nuclear crater which has a diameter of 400 metres. Note the numerous secondary impact craters round the main crater.

Chapter Seven

A Satellite's View of the Earth

K. FEA

Introduction

One of the most significant advances made, mostly incidentally, in the twentieth century concerns our view of the Earth. Understanding of the structure of the Earth has increased steadily since the late nineteenth century, but the achievements in the field of geophysics have resulted largely from the activities of what is loosely termed the 'Space Age'. That detailed information about the Earth—and its interior—should derive from experiments with sounding rockets and orbiting satellites seems a little odd. The reasons are: the Earth's gravitational field at distances of a few hundred kilometres is less disturbed by local irregularities than is the field close to the surface; the orbital motions of Earth satellites provide very sensitive indications of the large-scale features of the gravitational field; these large-scale features relate to the departures of the Earth's interior from spherical symmetry; large-scale structures in the surface morphology of the Earth become apparent in the photographs and television pictures obtained at rocket and satellite altitudes; similarly, large-scale aspects of the Earth's atmosphere become observable—for example, the cloud distributions over oceanic and continental areas indicate mesospheric processes on a global scale.

Speaking very generally one may include non-optical 'pictures' of the Earth under the term 'views of the Earth from space'. Infra-red emission variations over the Earth's surface have, for example, provided powerful levers on the problems of climatology and large-scale meteorology. The data from narrow-field infra-red detectors, if obtained with a high enough density, build into maps which have immediate visual impact. Indeed infra-red pictures of high quality have been returned from the Nimbus satellites and other similar vehicles. However, here we will restrict ourselves to conventional pictures acquired from spacecraft with straightforward cameras. The camera need not be hand-held; automatic cameras in unmanned recoverable spacecraft have yielded large numbers of excellent photographs of the Earth from various distances, but the more open 'programming' possible with manned systems has produced greater numbers of very high quality photographs from distances of about 100 km to about the distance of the Moon.

To illustrate the general appearance of the Earth from orbital altitudes Plate XXX reproduces a frame from the Mercury 9 flight of 1963. With most photographs from moderate altitudes the only features readily seen are complex cloud structures, a general blue haze and an indistinct horizon. Usually, unless the geographical position of the picture centre is stated, it is not even possible to determine if land or sea covers that part of the surface.

For comparison a frame from the Apollo 12 flight of 1969 is shown in Plate XXXI. The Moon is photographed from an altitude similar to that of the spacecraft from which Plate XXX was obtained—that is, about 200 km. The contrast is striking. The clarity of the lunar surface, resulting from the absence of a scattering atmosphere, allows details much smaller than 1 km to be seen with high contrast, even at the horizon. The horizon is sharp but relatively close. The curvature of the horizon is

Plate XXXIII
Comparison of albedoes of Earth and Moon; photograph from Apollo 10 spacecraft in orbit over lunar surface. *NASA photograph.*

Plate XXXIV
Photograph of the Earth from spacecraft en route to the Moon; Africa; Indian Ocean on the terminator. *NASA Apollo 11 photograph.*

Plate XXXV
Photograph of weather circulation system over Pacific. *NASA Apollo 10 photograph.*

Plate XXXVI
Cloud sheets with wave structures; altitude 150 km. *NASA Mercury 9 photograph.*

Plate XXXVII
Wave clouds and ice-covered lakes in Tibet; altitude 150 km. *NASA Mercury 9 photograph.*

Plate XXXVIII
Eddy formation in cirrus clouds off the West coast of Africa. *NASA Gemini 5 photograph.*

Plate XXXIX
Photograph of the Earth eclipsing the Sun. *NASA Apollo 12 photograph.*

Plate XL
South-western Tibet; large ice-covered lakes are Rakas Tal and Manasarowar; mountain mass shows snow-filled valleys. *NASA Mercury 9 photograph.*

greater; the Moon's radius being about one-quarter that of the Earth, the curvature of the horizon is four times that of the Earth's horizon for a given altitude. For the Earth an altitude of 200 km gives a distance to the horizon of about 1,500 km; the same altitude above the lunar surface would give a horizon approximately 750 km distant.

Further comparison of Earth and Moon is given in Plate XXXII. Reproduced to give the relative diameters in correct proportion the two photographs show the Earth and Moon from great distances. The picture of the Earth was obtained during the 1969 Apollo 10 flight, from a point about midway between Earth and Moon. The picture of the Moon, in an unusual aspect from a terrestrial observer's point of view, was obtained during the return flight of Apollo 11. The colours are roughly correct, though various factors cause the photographs to show spurious colours at some brightness levels. Note that the pictures seem to indicate equivalent surface brightness for the two planets. This is a result of exposure control. The photographs were given exposures to give normal registration of the images, so that the much duller lunar surface appears comparable in brightness to that of the clouded Earth.

The ALBEDO of the Earth, as a planet, is high. It reflects on average about 40% of the incident light. Densely clouded regions may reflect as much as 50% of the incident sunlight and fresh snow fields nearly 60%. The land and cloud-free sea areas reflect less; the lowest albedo values are found for vegetation-free land areas and volcanic fields, where the albedo falls to lunar values below 10%. Note, however, that dark rocks if finely divided will have a relatively high albedo, for example sandy and dusty deserts.

The Earth, like the Moon, exhibits marked backscatter in the reflected light. Clouds will appear significantly brighter at small phase angles, as any high-flying aircraft passenger will have noticed when looking at cloud sheets below and in directions opposite to that of the Sun.

SPECULAR REFLECTION does not occur on the Moon (excluding the reflections from polished man-made objects abandoned on the lunar surface). The Earth does sometimes show obvious specular reflections of the Sun. In Plate 66, six photographs of the Earth are placed in a series with increasing phase. The pictures are line-scan television pictures transmitted by the first of the ATS (Applications Technology Satellites) principally designed for communications and Earth's cloud-cover experiments. Bright areas are seen at all phases, the smeared image of the Sun moving towards the centre of the disk as the Earth approaches 'full Earth'. The intensity of the reflection varies; if the image lies in an ocean region the brightness is high, especially if the region is cloud-free and the sea calm. In fact the angular spread of the image indicates the roughness of the sea. A dead calm would permit a solar image of very high brightness —a brilliant star. Almost always the specular reflection is spread over some $10°$ of longitude and largely obscured.

Plate 67 completes the phases sequence with television pictures transmitted to Earth from Surveyor 7 on the Moon's surface.

The difference in the albedoes of the Earth and the Moon is shown very clearly in photographs which portray both disks in a single frame. Recently there have been quite a number of such pictures, particularly from the Apollo flights. These photographs, especially those showing earthrise or earthset at the Moon's limb, have considerable dramatic value. Plate XXXIII illustrates one of these photographs, obtained from the Apollo 10 flight of 1969. The illuminated half of Earth is just clear of the lunar horizon; the exposure is roughly correct for the Earth but the lunar surface is very under-exposed. The exposure for the Earth's distant disk is about that required on the Earth's surface, for surface brightness is independent of distance.

The colour and albedo of the Earth have of course been known with limited accuracy for many years. The phenomenon of EARTHSHINE on the night side of the visible disk of the Moon allows the properties of reflected light from the Earth to be studied. This is the well-known effect described as the 'old Moon in the new Moon's arms'. Some information on the gross albedo of the Earth and its variations with phase and time were obtained in this way long before photographs from space vehicles became available. More recently, for example on the Mariner II flight to Venus in 1962, the brightness variations of the integrated light from the Earth have been studied with photodetectors associated with Earth-sensor orientation systems. The telemetered data show that the day-to-day variations of the Earth's brightness can be as much as a factor of two, nearly one STELLAR

The Earth and Its Satellite

MAGNITUDE. Unlike Venus which in the Earth's sky shows large variation with phase but no significant variations due to cloud-structure variations, the Earth would show obvious fluctuations in brightness if observed from Venus. The rotation of the Earth and the short time-scale of the global weather patterns produce brightness variations of nearly one magnitude over periods as short as 24 hours.

Uniqueness of Earth's Atmosphere

The Earth is peculiar in the solar system in having an atmosphere partly transparent and partly clouded. Mercury and the Moon possess thin and perfectly transparent exospheres only. Venus possesses a thick cloud layer in its very dense atmosphere; something like 50 km of clouds prevents any albedo changes relieving a monotonous white disk. The JOVIAN PLANETS show latitude variations of brightness which indicate zonal circulation, and spots with comma-like tails suggesting vortex systems analogous to the low-pressure systems on Earth, possibly similar in origin to the polar front systems. Only Mars shows a transparent atmosphere with changing clouds, but the lower atmosphere of Mars has only about 1% of the Earth's tropospheric density. However, the clouds of Mars, both

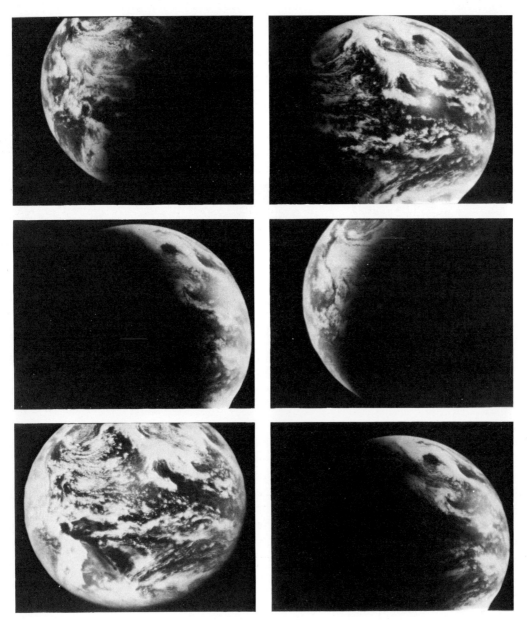

Plate 66
ATS-1 satellite television pictures showing the Earth passing through its phases; altitude 35,000 km; December 1966.
NASA photograph.

A Satellite's View of the Earth

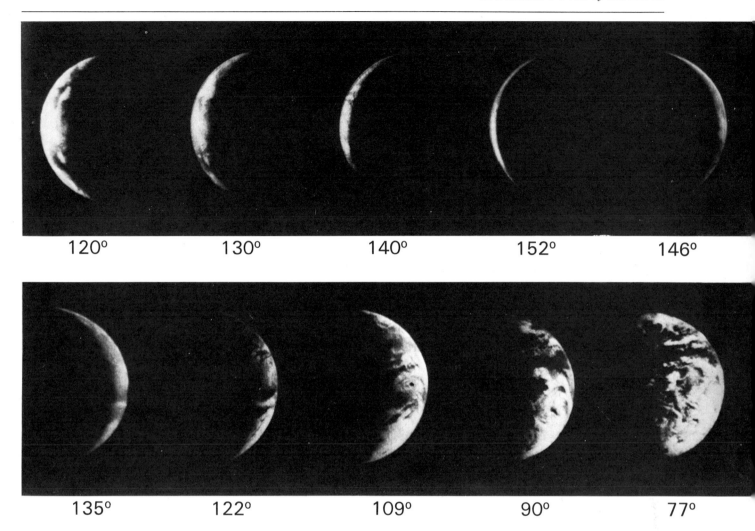

Plate 67   Surveyor 7 sequence of television pictures of Earth from the Moon's surface; January 1968. *NASA photograph*.

white and yellow (dust), can give an outline of the circulation systems in the lower atmosphere of Mars. Only the Earth shows extensive cloud systems which permit detailed investigation of the global and local processes of the troposphere and lower stratosphere. It is from this point of view that the pictures from spacecraft have such value, not only in relation to research but in relation to general education. Climate and weather features are immediately appreciated when the pictures are examined, even cursorily.

Plates 68 and XXXIV both show the Earth in GIBBOUS PHASE from distances of about 100,000 km. At this distance the Earth presents a full hemisphere, the angle between the centre of the disk and the limb being about 87° (geocentric). A number of aspects are very clear.

In both photographs the land mass of Africa shows distinctly, with cloud systems concealing the land details at higher latitudes. The red colour of the deserts in the Sahara and Kalahari regions is striking. The overall colour of the rest of the disk is blue. This strong blue colour of the Earth has been remarked by all the astronauts. It makes the Earth again outstanding among the planets. From the Earth, Mars appears orange; to the practised eye Mars justifies its bloody mythological associations. But the Earth from space shows a blue colour more obvious than the red of Mars. The reason is largely the scattering properties of the lower atmosphere. The same molecular scattering process that causes the clear terrestrial sky to be blue also causes the blue haze which covers the Earth when viewed from space.

Plate 68 Photograph from automatic camera on board Russian spacecraft Zond 5; Earth seen from 90,000 km during return flight from around the Moon.
Novosti photograph.

Plates 68 and XXXIV also show very clearly the global distribution of the weather patterns. Pictures of this sort could usefully find a place in elementary textbooks on physical geography, for the large-scale features of the climate as well as the dispositions of land masses of various types are immediately seen.

Smoothing out the details of the local weather systems, the global atmospheric circulation becomes apparent. The INSOLATION is greatest at the sub-solar latitudes, that is in the tropics. The surface heating reaches a maximum near the latitude over which the Sun stands: the equator at the EQUINOXES, the tropics at the SOLSTICES. One expects the strongly heated air to generate massive vertical currents and the immense equatorial cumulo-nimbus clouds to develop. Viewed from space, these show as a ragged band along the equator and this white band is an obvious feature in Plates 68 and XXXIV. The massed clouds grow during the morning and reach a maximum in the middle afternoon, decaying in the evening to some extent. Plate XXXIV includes the sunset TERMINATOR east of Africa over the Indian Ocean. Plate 68 on the other hand shows the sunrise terminator lying over the Atlantic Ocean west of Africa. Differences in the appearance of the equatorial clouds in the morning and afternoon zones are apparent. But superimposed on the local time variations are differences due to positions of land and sea. The north-east and south-east trade winds blow moist air from the Indian Ocean over the African continent; the air along the western coasts of Africa is, on the contrary, dried from its traverse of the land. These factors affect the cloud distributions and can be seen in the photographs.

While the air in the equatorial region ascends, driven by the solar heat engine, the air in the middle latitudes is descending. Therefore the equatorial cloud belt is bordered by relatively cloud-free zones, which again show clearly in the photographs. The meridional flow, which is directed towards the equator at low levels, develops the familiar zonal flow of the trades under the action of the geostrophic forces (CORIOLIS FORCES). At higher latitudes the low-pressure systems develop where the warm air from low latitudes meets the polar fronts.

The depressions so familiar at latitudes between about 40° and 60° are generated when warm, and commonly maritime, air encounters cold polar air during eastward and poleward motion. Instabilities with dimensions of about 1,000 km develop, producing areas of lower pressure at low levels. The subsequent slow filling causes inflowing air to circulate with the geostrophic winds that sweep around the low-pressure centre in an anti-clockwise direction in the northern hemisphere and in a clockwise direction in the southern hemisphere. The sense of the rotation is easily seen in the photographs, Plates 68 and XXXIV. The Zond 5 photograph shows particularly well the opposite rotations in the regions farthest north and south near the terminator. Near the most northern part of the limb and terminator two related low-pressure regions are clearly linked. This is a characteristic of mid-latitude weather; depressions may follow one another like beads on a necklace.

Plate XXXV shows more details of a complex circulation pattern over the north Pacific, a feature which shows also in Plate XXXII. This photograph also was taken during the early part of the Apollo 10 flight but the detailed view reveals the interaction of at least two streams in the lower atmosphere. The sinuous cloud forms often seen snaking over ocean regions are not well understood. Sometimes they appear related to large-scale flow patterns which are interacting; sometimes they appear to follow the directions of the ocean currents below. From the time the first photographs were obtained from rocket altitudes the large-scale organisation of clouds over distances upward of 500 km has been evident. Wave clouds in the lee of mountains are very obvious in many of the pictures, with wave-like disturbances stretching for hundreds of kilometres downwind of extended mountain ridges. Cellular patterns similar in form to the patterns often seen from the ground in alto-cumulus and cirro-cumulus sheets are also observed from satellite altitudes. But the extent of the patterns turns out to be surprisingly large; since the patterns are often observed over ocean areas they cannot be due to orographic effects. Some insight into the physical processes underlying these large-scale cloud organisations may be gained from infra-red studies, also from satellite altitudes. Such studies have been carried out extensively by the Tiros, Essa and Nimbus vehicles over the past 10 years.

Two examples of wave-forms in cloud sheets are shown in Plates XXXVI and XXXVII. Both pictures were obtained on the Mercury 9 flight of 1963. This flight was the first manned orbital flight to include a serious terrestrial photography programme. Subsequently the Gemini manned flights developed the synoptic terrain and weather photography programmes, and many hundreds of high-quality colour photographs were obtained. The cameras usually are Hasselblad systems with frame sizes of 70 mm format. Conventional 35 mm cameras have been used also on most missions but return pictures of lower resolution. In the Apollo flights which have followed the Gemini flights, synoptic terrain photography has been included in some of the earlier development missions. The Apollo 9 Earth orbital flight, in particular, resulted in the acquisition of some pictures of geographical and geological structures of unprecedented quality.

The value of satellite surveillance of cloud distributions is well known. Many times in recent years weather satellites have provided the first indications of cyclone and hurricane formations growing over the ocean regions, especially in the Atlantic and Indian Oceans. Adequate coverage of weather phenomena to aid forecasting and hurricane prediction can only be obtained with continuous transmissions from operational weather satellite systems. The value of high-quality photographs obtained during manned flights is mainly in the areas of geology and atmospheric studies of a fundamental nature. Understanding of lower atmosphere phenomena is advanced, for example, by the pictures showing eddy structures in cloud sheets and spiral forms related to surface features, such as islands in the sea and mountains on land. Plate XXXVIII illustrates vortex-like structures seen during the Gemini 4 and 5 flights of 1965. On the smallest scale, though still large compared with the view from the Earth's surface, the spiral forms do not necessarily adhere to the rotation law for opposite hemispheres. In some cases rotating spiral eddies form with adjacent spirals rotating in opposite senses, clearly influencing one another. This is a mechanism distinct from that producing the spiral forms of low-pressure weather systems. In this latter case the spiral forms are generated when inflowing air tending to fill a low-pressure region deviates under the inertial forces arising from motion on a rotating planet.

Surveillance Photography from Orbital Altitudes

Below the almost universal clouds over the Earth's surface the details of the surface features are not easily registered. The majority of the photographs taken from orbital altitudes show only cloud formations. Even in cloud-free areas the inevitable haze obscures fine details by lowering contrast—an effect present in vertical views but especially obvious in oblique views. The details at the horizon are always hidden unless infra-red photography is employed. Very little use has been made of infra-red photography since the Viking 11 and 12 flights to about 250 km over White Sands, New Mexico, in 1954–5. Then aerial reconnaissance cameras (K-25 type) were used and infra-red films and filters gave high resolution of the surface details even at the horizon, 1,500 km distant. Analysis of the Viking 11 photographs showed that the resolution was of the order of 200 metres from 250 km altitude. This was based on the visibility of fine linear features such as airfield runways and railroads.

Naturally the visibility of details depends on a number of factors. The three most important factors, assuming adequate resolution capability of the vehicle instrumentation, are contrast, linear extent and elevation of the Sun. That resolutions greatly exceeding the Viking results are now possible can be assumed since 16 years have advanced the state of the art in surveillance photography; reading between the lines in accounts of satellite (military) reconnaissance it is clear that objects of the order of 10-metre size can be revealed from photographs obtained by automatic satellites (probably significantly better than this figure for very favourable situations); the success of the lunar Orbiter spacecraft, using photographic systems engineered by Eastman Kodak, and which are presumably based on the systems carried in the military reconnaissance satellites, suggests that under ideal conditions it may be possible to detect surface details on the Earth of the order of 2-metre size, i.e. a man at sunset. However, such extreme resolution is not really necessary for geological studies. For in geological studies from space vehicles the greatest value attaches to the coverage of large regions in single photographs. Such pictures can be obtained without space vehicles only if very comprehensive aeroplane photography is organised, with the resultant photographs subsequently being mosaicked together. But then the operation is costly and difficult, and even quality over large regions is difficult to achieve.

The appearance of the lower atmosphere at the

Plate 69   Apollo 7; Atlantic coasts of north-eastern Uruguay and southern Brazil; Lagoa Mirim and Lagoa dos Patos. October 1968. *NASA photograph.*

horizon has already been noted. A distinct band occurs at the horizon, with an apparent thickness about equal to the Moon's diameter, or half a degree. A simple calculation shows that the half-degree subtended by the opaque atmosphere at the horizon 1,500 km distant implies a thickness of about 15 km for the absorbing and scattering troposphere. In fact, at middle latitudes, this is about the accepted height of the tropopause. Convection to the tropopause will maintain enough dust in suspension to give rise to the strong obscuration in the visible wavelengths. Infra-red wavelengths are much less affected.

A very dramatic illustration of the scattering properties of the Earth's atmosphere is given by the photograph of the Earth eclipsing the Sun, Plate XXXIX. During the return flight of the Apollo 12 spacecraft from the Moon the astronauts observed the total eclipse of the Sun by the Earth. In this photograph the Sun is just disappearing behind the Earth's limb but the lower atmosphere strongly scatters all wavelengths. So a white arc surrounds half the Earth. At the cusps coloured beads appear, the points farthest from the Sun (in geocentric angular distance) showing red. The blue light is scattered most strongly, allowing the red light to illuminate the atmosphere beyond the point where the blue light is extinguished. A similar argument explains the red illumination of the totally eclipsed Moon sometimes seen at mid-lunar eclipse.

Plate 70 Apollo 7; Taumotu archipelago in South Pacific; altitude 200 km. October 1968. *NASA photograph.*

In cloud-free areas some interesting features of the Earth's surface appear, especially in near-vertical views. It turns out to be unexpectedly difficult to distinguish coastal features from submarine features in the shallow waters near the coasts. This is particularly true for the ocean atolls, where numerous underwater structures seem, from orbital altitude, to be islands. Plate 69, a photograph taken during the Apollo 7 flight (Earth orbital; altitude approximately 300 km), shows the coast of north-eastern Uruguay and southern Brazil. Two lagoons lie inland of the coastal bar and the shallow water in the lagoons allows details below to be seen. Off the coast underwater details also can be clearly seen up to the point where the sea bottom dips more steeply.

In Plate 69 the coast itself is defined by the bright sandy beaches. Finely divided sand, as remarked earlier, is highly reflective and white in colour. The bright lines of wide beaches serve as positive aids in following coastlines on satellite photographs. In Plate 70, also obtained on the Apollo 7 flight, the skeleton outlines of the islands in the Tuamotu archipelago in the Pacific are made visible by the white sand beaches.

Larger islands, and land masses in general, are often overlain by massed cumulus clouds. While these cloud covers prevent the surface details being seen they provide an easy key to the identification of the land areas. The cumulus clouds develop over the more strongly heated land and their concentration ends at the coasts. Sometimes the lines of

Plate 71   Gemini 11; photograph of India and Ceylon; Agena transponder in the foreground; altitude 850 km. September 1966. *NASA photograph.*

clouds appear to commence some distance inland, but following closely the lines of the coasts. Sea breezes, blowing from the sea inland, cool the coastal strip and also blow the clouds back from the coast. A number of these features are visible in Plate 71. This impressive picture was taken during the Gemini 11 flight; the photograph shows most of India and Ceylon from a height of about 850 km. (The much greater height attained in the Gemini 11 mission resulted from the successful use of the power of the attached Agena rocket, with which the Gemini spacecraft made a rendezvous in orbit and subsequently hard docked.) In this photograph Ceylon is outlined in cumulus clouds. Similarly the western coastal regions of India are also cloud-covered. Clouds of various types thinly cover most of India, making the shape of the country obvious but land details obscure; note the curious patterns of broken clouds around the bordering seas (the processes generating these patterns are not well understood).

A New Look at Cultivated Areas

Two other features characteristic of the planet Earth are shown outstandingly well in the orbital photographs. Rivers, both active and dry, show clearly wherever they wind across extensive areas. Dry river-beds frequently appear as bright as reflecting water, for both fine sand and salt have high albedos. Plates 72 and 73 illustrate two great river regions. The Ganges Plain lying south of the

Plate 72 Apollo 7; view looking east across the Ganges plain; Himalaya mountains to the north; Patna near joining of rivers. October 1968. *NASA photograph.*

hills of Nepal is crossed by numerous rivers; the Ganges is joined by the Ghaghara and the Gandak, near the city of Patna. All the channels, even of the small tributaries, can be readily traced in detail. In Plate 73 the Mississippi River valley is similarly portrayed. As well as numerous tributaries some indications of the cultivated areas are visible. Small squares merge into bright masses which represent intensive agricultural development. Recently, particularly on the Apollo 9 flight, experiments were made with infra-red colour film techniques. It has been shown that a great deal of information can be obtained from orbital photography on the state of cultivated areas. It appears that healthy and diseased crops have different infra-red and visible region albedoes, permitting large-scale studies of the vegetation state. Similarly, a great deal of attention has been given to the question of the detection of natural resources by satellite surveillance methods. Much optimism has been expressed on the possibilities of locating new mineral- and oil-fields from geological mapping based on satellite photographs.

The problems of mapping in general, but especially of geological mapping, are partly solved with orbital techniques. Inaccessible regions can be studied in great detail and the methods of photogrammetry applied to obtain contours, provided the photographs are properly oriented and, prefer-

Plate 73 Apollo 7; view of Mississippi river valley looking north; parts of Louisiana and Mississippi; altitude 190 km. October 1968. *NASA photograph.*

ably, taken in stereo pairs. Mountainous regions can be explored geologically without the hazards and discomforts of physical exploration (though, if the photographs are obtained by manned spacecraft some other hazards and discomforts are involved). Plate XL illustrates a mountain massif in the Himalayas; the picture is one of the finest obtained on the Mercury 9 flight. The snow-covered ridges and valleys are immediately visible and some details of the ice-covered lakes at lower altitudes are seen. The general high elevation of the Himalayan region aids visibility; the dust is confined to low levels and the air over the mountains is very transparent. This photograph can be compared with the photographs of lunar mountain areas elsewhere in this book. The differences are immediately apparent. Apart from the deep erosion channels which carve the slopes of all terrestrial mountains, the appearance of the Himalayan range is quite unlike the great lunar mountain ranges.

The wealth of information in the photographs of the Earth from satellite altitudes has had an impact on many geophysical fields. The impact on man's view of the planet he lives on is less obvious. But the impact is considerable. From the time Anaximander speculated on the Earth's suspension in the void it has taken nearly 2,500 years to obtain these views of the Earth isolated in space.

Chapter Eight

Surface of the Moon

E. L. G. BOWELL

The Moon is easily the brightest celestial object visible in the night sky. Its greenish-white light is sufficiently bright to read by—at least 4,000 times brighter than its nearest celestial rival, Venus, and 60,000 times more luminous than the brightest star, Sirius.

Together with the Sun it is the only astronomical body which displays a disk to the unaided eye; even casual observation reveals a large number of surface features (*Plate* 74A).

Earth-based Observations

Our eyes are sufficiently good optical instruments to resolve details as small as one-thirtieth of the Moon's diameter, representing a distance of 150 km on the lunar surface. We can easily discern the boundaries between MARIA and HIGHLANDS, and just detect some of the largest craters.

Using binoculars, or a telescope of moderate aperture, the number of details visible is increased dramatically. For example, binoculars magnifying ten times can isolate features as small as 10 km diameter. About 5,000 craters will be brought into view, along with many features of other kinds: RILLES, WRINKLE RIDGES (see MARE RIDGE, and the central peaks of some craters.

Larger instruments increase the number of visible features still further. A telescope of 1 metre aperture reveals, in the best observing conditions, craters less than $\frac{1}{2}$ km in diameter (*Plates* 74B, 75A). However, this is the limit of the detail which Earth-based telescopes can show. The use of still larger telescopes or high magnifications cannot combat the unsteadiness of the lunar image caused by the turbulence of the Earth's atmosphere.

The amount of detail seen by the eye through the telescope is usually greater than that recorded on the photographic plate. Motion of the atmosphere has the effect of blurring the image during a photographic exposure, while the eye can often follow the trembling motion of a tiny lunar feature.

One other hindrance annoys the Earth-based observer. From Earth we can view only part of the Moon's surface, and then from only one angle. Near the edge of the Moon's disk (the LIMB) the topography is always seen much foreshortened. Circular craters are distorted into elliptical shapes; a mountainous region might completely hide a lower-lying plain. To study these limb regions, photographs of the Moon are often *rectified*: reprojected so that the surface appears as though viewed directly from above (*Plate* 76).

It was pointed out in Chapter Two that 59% of the Moon's total surface area can be seen; the remaining 41% is perpetually hidden. To see the far side of the Moon and to improve the image quality still further it is necessary to leave the Earth.

Observation from Orbiting Space Vehicles

Images of the Moon's surface have been obtained both televisually and photographically from unmanned space vehicles. Manned missions have produced pictures of higher quality than the automatic stations, but only for certain carefully chosen lunar sites.

To compare image quality it is interesting to study the photographs relayed from the American Orbiter series of missions in 1966–7. These vehicles,

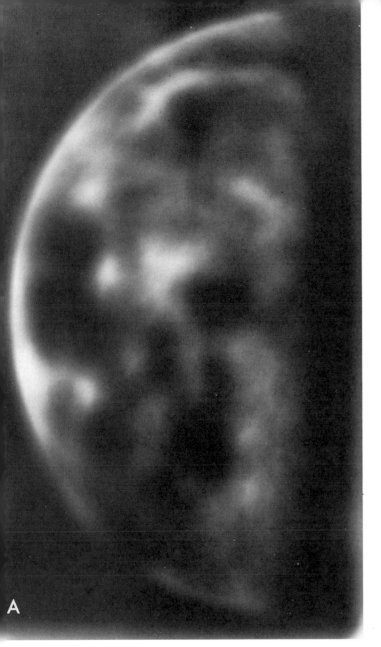 

Plate 74A Simulated naked-eye view of the Moon at last quarter. The eye can see objects as small as 150 km across (see scale on (B)). The outlines of maria and highlands are easily discernible.

Plate 74B Large telescope view. Resolution 800 metres. Inset locates Plate 75(A). The bar at lower left indicates a distance of about 500 km.

five in all, were placed in orbit around the Moon, either about its equator (Orbiters I, II, III) or over its poles (Orbiters IV, V). Each craft carried two cameras for medium- and high-resolution pictures; and in all 95% of the Moon's entire surface was photographed.

A medium-resolution shot of the Aristarchus plateau (*Plate* 75B) should be compared with the best obtainable Earth-based photograph (*Plate* 75A). Very many more details are visible on the Orbiter photograph. The next photograph in the series (*Plate* 75C), is a closer view of the crater Aristarchus taken by Orbiter V. Features as small as 40 metres across can be picked out; this is ten times better than attained by Earth-based observations. Finally, Plate 75D shows part of a high-resolution Orbiter V frame, centred near the mountainous hummock in the floor of Aristarchus. The boulders which can be seen scattered on the relatively smooth slope of this hummock range in size from about 30 metres down to 2 metres across.

Photography performed during the Apollo missions provides the highest resolution of the lunar surface so far obtained from orbiting craft. The 70 mm frames were returned to Earth for processing, and show details as fine as $\frac{1}{2}$ metre diameter. Many

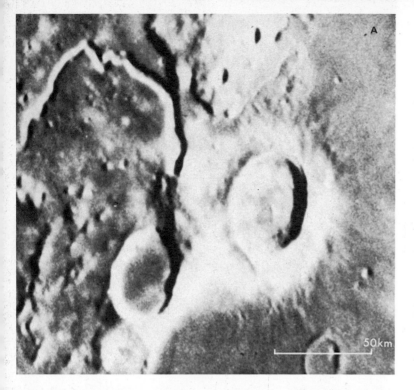

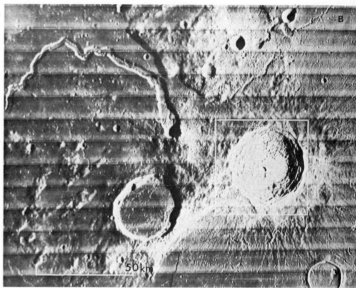

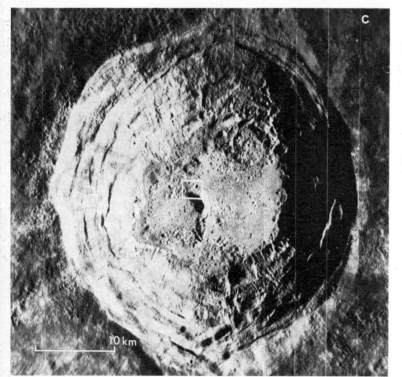

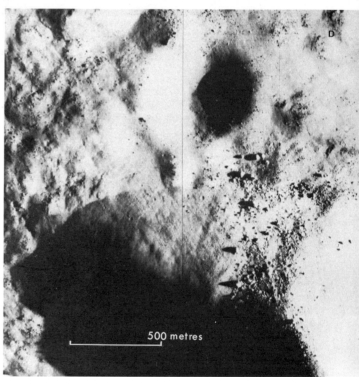

Plate 75A Aristarchus plateau. Best Earth-based telescopic view. Resolution 400 metres.

Plate 75B As (A) Photograph relayed from Lunar Orbiter IV. Resolution 100 metres. Inset locates (C).

Plate 75C Orbiter V medium-resolution shot. Resolution 40 metres—ten times better image quality than (A). The large crater, Aristarchus, is 40 km across. Inset locates (D).

Plate 75D Orbiter V, high resolution. The hummock near the centre of Aristarchus' floor is strewn with boulders, some as large as 30 metres diameter. The smallest visible has a diameter of about 2 metres. This is one of the highest resolution shots telemetered to Earth. *NASA photograph.*

Plate 76A (*opposite page*) Earth-based view of the western lunar limb in the region of the crater Struve and Eddington.

Plate 76B Rectified image of the same region. The cluster of craters on the left is situated at lunar longitude 87° W. West is at the top.

of these photographs were taken for the reconnaissance of specific sites to determine their suitability for manned landing.

Lighting Conditions

The monthly rotation of the Moon about its axis in 27·3 days causes each point on the surface to be lit for a little under 14 days. Just as it does on Earth, the Sun rises in the east and sets in the west, illuminating craters and plains from different angles during the course of the day. From the Earth the appearance of a crater or peak can change surprisingly. To illustrate some of the effects of different lighting conditions Plates 77 to 79 show three views of the Aristarchus plateau.

Even more startling lighting changes can be seen in Plate 80, which shows two unnamed far-side craters near the large crater Tsiolkovsky. The smaller of these is 20 km across. In view B the Sun is only 13° above the local horizon, and the shadows cast by the crater rims extend well across the crater floors, enabling small raised or depressed features to be picked out easily. In view A, taken with the Sun at an altitude of 69°, it is quite difficult to recognise many of the features. The slopes in the terrain to the north of the craters have disappeared entirely, and even the two principal craters are visible only by virtue of their bright rim material.

Why do some but not all of the small craters shown in the high Sun view appear as bright points of light? There is a pair of similar looking craters (each about 4 km across) to the north of the two largest craters which can be well seen in the low Sun view (B) that almost disappear in the high Sun view (A). Nearby, just north of the 20 km crater, is a very tiny crater, insignificant on the low Sun view but very prominent at a high Sun angle. The general rule is that craters with sharply defined outlines show up brightly at high Sun, while those with softened rims fade into the background. Almost certainly most of the craters with rounded rims are old features that have been greatly eroded. The craters with sharp rims are of relatively recent origin. As a first approximation the brightness of these small craters is a measure of their age: the brighter they appear the younger they are.

A look at the full Moon disk in Plate 81 shows a number of very bright craters, usually accompanied by extensive systems of *rays* radiating from their centres. These craters, such as Tycho, Copernicus,

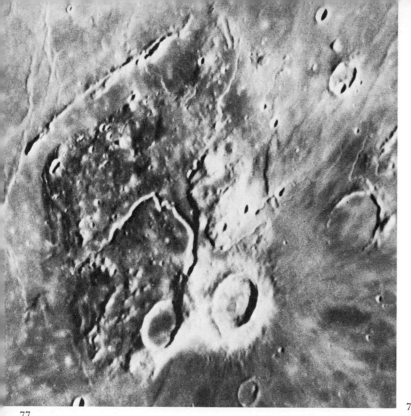

Plates 77 to 79 Three telescopic views of the Aristarchus plateau with different lighting conditions. The three largest craters are (*left to right*) Herodotus, Aristarchus, Prinz (part of whose ringwell is missing). *Consolidated Lunar Atlas.*

Plate 77 Morning view. The whole plateau stands out by virtue of long shadows.

Plate 78 Noon (high Sun) view. The effect of relief is lost, but the Aristarchus ray system shows up well.

Plate 79 Evening view. The three-dimensional aspect is reinstated, but opposite lighting results in this view looking almost like a negative of 77.

Kepler and Aristarchus, are known to be very young features.

At low Sun angle it is possible to discern features raised only a few metres above the general level of the surface by the long shadows they cast. Small ridges and FLOW FRONTS in the maria can be detected in this way. Measuring shadow lengths provides a means of determining the relative heights of mountains and ridges, and the shapes of crater interiors can be deduced from a study of the changing aspect of shadows cast on their floors.

The Colour of the Lunar Surface

The eye sees the Moon as having a greenish-white hue, but careful measurement of the brightness of the Moon through colour filters shows it to have a slight overall reddish coloration.

Differences in colour exist also, and these can be shown to great effect by an ingenious technique. Two monochrome photographs of the Moon are taken: one with a red filter over the photographic plate, one with a blue filter. When processed these are both negatives on which light areas appear dark and vice versa. Next, a positive is made of the blue-filter image, so that this is in effect a normal view of the Moon in blue light. Then the red-filter negative

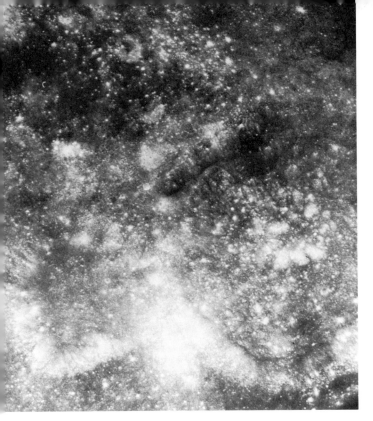 

Plate 80 Views of two unnamed craters, the smaller of which is 20 km across. (A) Apollo 8 picture. The Sun angle is 69°. Relief is completely washed out, except in some crater rims. Note that many small craters, inconspicuous in (B) now show up very prominently as bright points. (B) Orbiter I photograph. The Sun is only 13° above the local horizon. Long shadows identify relief, particularly on slopes at the top of the picture. *NASA photographs.*

and blue-filter positive are superimposed. The effect is to exaggerate the colours of lunar surface features. Blue features appear as light markings and red features as dark markings.

Plate 82 shows an image of the Moon almost at full phase obtained in this way. Most surprisingly, the tones are not very different from the ones we are used to. Overall, the image seems like a negative of a typical full Moon photograph. The reason for this effect is that bright areas (the highlands, young craters and ray systems) are generally redder than dark areas (the maria, some old crater floors). Subtle colour differences in the maria enable geologists to work out the history and mechanism of formation of these regions.

Lunar Gravity

The pull of gravity on the Moon is about one-sixth that on the Earth. In figures, the acceleration is 162 cm/sec^2 compared with the terrestrial value of 981 cm/sec^2. This means that any object projected from the Moon's surface falls to the ground much more slowly and travels a greater distance than it would if impelled at a comparable speed on the Earth. A game of lunar golf would be an intriguing experience. The ball could be driven about 4 km, rising about a kilometre above the ground, and its flight would last over a minute, taking it out of sight over the horizon. The lack of a lunar atmosphere—and therefore winds of any kind—might make the game boringly repetitive, but the profusion of natural bunkers would certainly bring out the golfers' skills.*

Any object on the Moon weighs six times less than on Earth. In particular, boulders, rocks and larger structures (ridges and crater rims) overlay the lunar surface with a smaller pressure. This is one reason why height differences as large as 10 km can be sustained for long periods. Another effect of low gravity is to increase the angles of slopes on which fine particles can rest without sliding.

Mascons

Studies of the orbits of Apollo and lunar Orbiter craft close to the Moon have revealed slight irregularities in the lunar gravitational field. Their cause has been attributed to mass concentrations, or *mascons*, beneath the lunar surface. When a space

---

* Captain Alan Shepard, Commander of the recent Apollo 14 mission, drove several golfballs over the lunar surface using a rock-collecting tool as a club.

Plate 81 Full Moon, Earth-based photograph showing many ray systems. *Consolidated Lunar Atlas.*

vehicle passes over a mascon it experiences an increase in the strength of the Moon's gravitational attraction that causes it to accelerate. Naturally, mascons are of very small mass compared with the Moon, and the accelerations they produce in the motions of spacecraft are correspondingly tiny: typically they are less than a thousandth of the Moon's steady gravitational pull.

After a number of spacecraft orbits the tracking data can be analysed to show the detailed positions of mascons on a lunar map (*Plate* 8). Surprisingly, most of the mascons are found in the circular MARIAL BASINS, such as Mare Imbrium and Mare Serenitatis. They do not occur with any prominence in the highland regions. One way to interpret this result is to imagine that dense bodies (perhaps metal-rich ASTEROIDS) hit the Moon, formed the circular maria and became buried beneath their surfaces. On this model the mascons are located at depths ranging from 50 to 210 km, and the largest of them has a mass equivalent to a 90 km diameter sphere of nickel-iron. However, the orbital tracking

Plate 82 Colour differences on the Moon: blue regions show up as light-toned and red regions as dark-toned. The maria are all bluer than the highlands, and the craters with ray systems are reddest of all. Compare the tones of the full Moon photograph in Plate 81.

data do not preclude the possibility that the mascons lie nearer the mare surfaces, in which case they can be represented as plate-shaped objects such as large volumes of dense igneous rocks. This is more likely to be the real situation.

Another way to look at the irregularities in the lunar gravitational field is to interpret the gravity contours as equivalent surface heights. In a region elevated above the average lunar surface there will be more underlying material and therefore the pull of gravity will be greater. The mascons in the circular maria now have to be seen as concentrations of material (and not mass), and the maria as regions elevated by as much as 3 km above the average lunar surface. This is plainly not the case, as a comparison between the heights shown in Plate 7 and the gravity contours (now supposed to represent heights) in Plate 8 shows.

More refined applications of orbital tracking data will eventually lead to planetary gravity mapping capable of identifying geological structures and showing the extent of mineral resources.

The Lunar Atmosphere

It is clear that the lunar environment is one of ultra-high vacuum, although up to the present time no direct pressure or composition measurements have been made. To get a sense of scale we can compare the estimated density of the lunar atmosphere with the Earth's. The sea-level pressure of air is 76 cm of mercury (known as 1 atmosphere pressure). The ground-level lunar atmospheric pressure is calculated at about 10^{-14} atmosphere. The most sophisticated vacuum systems so far devised in terrestrial laboratories are capable of maintaining pressures as low as 10^{-12} atmosphere only for short periods; the lunar atmosphere is 100 times more rarefied still. This very low gas pressure is a difficult figure to comprehend; it means that the Moon's entire atmosphere weighs only a few tons and could be stored at sea-level air pressure in a building about the size of a large house.

It is fairly easy to prove that the Moon is not enveloped by an extensive atmosphere. In its motion around the Earth the Moon often passes in front of stars (the phenomenon of *occultation*) and when it does so their light is instantaneously extinguished. This would not be the case if the Moon possessed an atmosphere even 1,000 times more rarefied than the Earth's. Occultation experiments on radio sources have shown the pressure to be less than 10^{-13} atmosphere.

Has the Moon ever had a significant atmosphere, and if so, what gases was it composed of and where did they go? The last part of this question can be answered fairly easily. A lunar atmosphere would slowly leak away into space because lunar gravity is not powerful enough to retain it. We can be sure that the light gases—hydrogen, helium—would escape from the Moon much faster than the heavy ones—carbon dioxide, krypton—so that a lunar atmosphere which started off as a mixture of a large number of gases would quickly become depleted of the lighter ones. During the Moon's lifetime all but the heaviest gases, krypton and xenon, would have leaked into space. Carbon dioxide should survive for about 2,000 million years, about half the age of the Moon.

Other arguments suggest that the lunar atmosphere is being continually replenished by gases from the lunar surface and from the *solar wind*. Once more the rules for the depletion of the atmosphere apply, but now if a light gas can be supplied sufficiently rapidly to keep up with its leakage rate it may be present in significant abundance in the overall composition. Most calculations show that the lunar atmosphere consists mainly of hydrogen, helium, water vapour and carbon dioxide, with some neon. Oxygen is not thought to be present in significant quantities.

As a consequence of low pressure, collisions between gas molecules occur much less frequently than they do in the Earth's atmosphere, perhaps once every two weeks, compared with 10^{10} times per second at sea level on Earth. Gas molecules therefore behave like ballistic particles above the lunar surface. A carbon-dioxide molecule, moving at nearly $\frac{1}{2}$ km per second, might 'jump' 100 km in seven minutes. The gas comprising the lunar atmosphere must be very mobile, moving rapidly over the whole of the Moon's surface, and colliding quite frequently with the surface at speeds of about $\frac{1}{2}$ km/sec.

With this view, it is interesting to predict what would happen if the lunar environment were contaminated, say by exhaust gases from a rocket—a very real problem in view of the small extent of the atmosphere. From the figures we have above, we can work out that carbon dioxide would take about ten hours to reach the opposite side of the Moon, making about 100 jumps on its journey. Gases from a rocket exhaust, such as water vapour and carbon dioxide, would therefore take only five to ten hours to contaminate the whole of the Moon.

Temperature

In some parts of the Moon the exposed REGOLITH (see Chapter Nine) is subject to temperature changes from 120 °C at lunar midday to −150 °C at midnight. Lack of atmospheric protection allows sunlight to strike the lunar surface with an intensity which we do not experience on Earth, where the circulation of the atmosphere and the presence of clouds and highly absorbent water vapour attenuate the Sun's heat during the day and retain it at night.

On the Moon heat is transported solely by radiation from the Sun (*insolation*), and cannot be spread by winds or absorbed by clouds and vegetation. The sunlit rocks attain temperatures as high as 300 °C at lunar noon. Heat can be dissipated only by conduction, but this is an inefficient process because lunar surface material is a very good insulator—so good that just a few metres below the surface the temperature does not vary at all, but stays constant at about −50°C. Figure 24B shows the variation of temperature for a point on the Moon's equator throughout a LUNATION, and also the change of temperature with depth in the regolith.

During an eclipse of the Moon, when the Sun's rays are cut off for two or three hours, very rapid temperature changes occur: a drop rate of 60 °C per minute and a rise rate of 45 °C per minute are typical (Figure 24A). These changes may be sufficient to cause some rocks to crack or throw off small chips, and therefore lunar eclipses may contribute to erosional processes. Lunation temperature variation is not thought to be rapid enough to cause major rock fragmentation.

An important phenomenon observed during the lunar night, and particularly during eclipses, is the presence of *hot spots*. When the Moon's surface is cooling rapidly at the onset of lunar eclipse, certain areas stand out as thermal enhancements. The temperatures in these hot spots may be as much as 30 °C higher than their environs. The probable cause is a high abundance of surface rocks. Calculations made for regions in the neighbourhood of Tycho indicate that the surface is 10% populated with large rock blocks.

Like the Earth there is an overall change of temperature with latitude. The lunar equator is subject to the highest temperatures and greatest changes. The lunar poles receive sunlight obliquely

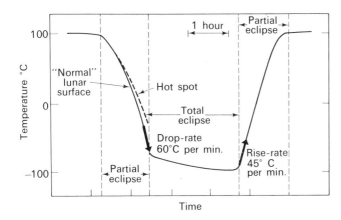

Figure 24A Temperature during a lunar eclipse for a point near the centre of the Moon's disk.

Figure 24B Temperature during a lunation for different depths (in cm) below the lunar surface.

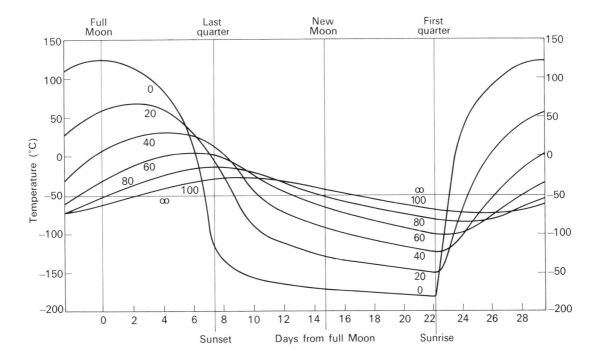

and probably have a fairly constant temperature below $-100\,°C$.

Ice under the Surface—Permafrost

Many scientists have argued that water once flowed over the lunar surface. The evidence for fluvial activity is slender, but it does raise the question of whether any water is still trapped in surface rocks or at depth. Obviously it will be very important to find water resources when manned lunar bases are established.

No water can exist in the hostile vacuum environment on the lunar surface—it would quickly vaporise and escape into space—but it might be found under the surface as ice. The term *permafrost* has been given to these possible deposits. Below most of the Moon's surface, at a depth of a few metres, the temperature is about $-30\,°C$. Promising sites ought to be at the lunar poles and near regions of volcanic activity.

The Moon has arctic and antarctic regions which experience polar days and nights of six-month duration. They are circular, about 100 km in diameter—much smaller than the polar regions on the Earth. The Sun never rises more than $1\frac{1}{2}°$ above the horizon at the poles, and so there are many nearby craters whose floors are permanently in shadow. Plate 83 shows this in three views of the lunar north pole with the solar illumination from different directions. Steep-sided craters outside the polar regions may also share this property—in all it is estimated that $\frac{1}{2}\%$ to 2% of the lunar surface is in permanent shadow.

This may be the location of most of the Moon's permafrost, which could exist in an extensive layer a few metres below the surface. The vestiges of the lunar atmosphere may also reside in the polar craters.

The Meteorite and Solar Wind Environment of the Moon

During three centuries of concentrated Moonwatching, no observation of a crater in the process of formation has ever been substantiated. And yet the Moon is densely pitted with craters, many of which must be the result of bombardment by rock debris originally in orbit around the Sun. Plate 9 shows the global abundance of lunar craters.

From the Earth *meteors* and METEORITES can be seen as 'shooting stars'. Counts of meteor trails, collection of particles by space probes and examination of stony and metallic meteorite fragments found on Earth show that about 1,000 tons of this material hits the Earth each day. A lesser quantity falls on the Moon because of its smaller size and gravitational pull (and the slight shielding effect of the Earth). The mean daily infall of meteorites on to the Moon probably amounts to 40 tons.

On average, meteorites hit the Moon at speeds near 15 km/sec; the slowest arrive at a little over lunar escape velocity (2·4 km/sec) and the fastest at about 30 km/sec—a considerable range. As a rule we can be sure that most meteorites (perhaps 99% by mass) are very small grains—*micrometeorites*. This is consistent with there being many more small craters than large ones on the lunar surface.

The size of crater which a meteorite makes on impact depends upon its mass and speed. Typically it will be brought to rest under the lunar surface at a depth of twice its diameter, and the energy generated as heat during the rapid braking will be dissipated in an explosion. Thousands of times a meteorite's mass can be displaced, much of this being thrown out, leaving a crater. The thrown-out material, if ejected rapidly enough, can subsequently form *secondary craters*, or can even escape from the Moon entirely.

To form a 1 km crater—a crater near the limits of detection for Earth-based observers—a meteorite weighing 40,000 tons impacting at 15 km/sec is required. Such massive objects, probably 20 metres across, very seldom encroach within the Earth–Moon environment. We would have to wait about three million years to see an event like this, so it is not surprising that no one has witnessed the birth of a lunar crater.

On a smaller scale, say for meteorites of 10 g mass upward, we should expect one impact per day in an area of 70,000 km², and still have to wait 30,000 years to see an impact in an area the size of a football field. The crater would be about $\frac{1}{2}$ metre across.

But when we consider the infall of micrometeorites the number of events increases dramatically. An area the size of a dinner-plate gets hit once per day by particles of 0·001 mm diameter or larger. These small grains, specks much finer than flour, are still capable of excavating craters 0·05 mm across (*Plates* 88, 89). On this microscopic scale the total damage done to the lunar surface is greater than that resulting from the more cataclysmic but much rarer event of a metre-sized meteorite impact. The great numbers of micrometeorites are sufficient to 'churn' the whole regolith in a period of 40 million years, and this makes them the main erosive force which wears down the sharp rims of fresh craters and slowly breaks up boulders and solid rocks.

Crater counts on the nearside of the Moon have shown the highlands to be up 30 times more heavily cratered than the maria (*Plate* 9). Presumably this means that the maria are much younger features but, additionally, the Moon must have been subjected to a much higher impact rate by meteoritic material in its early history—mainly during the first seventh of its life. The larger numbers of craters that characterise the lunar far side could have come about when the Moon was much closer to the Earth.

The Moon is hit by even smaller particles than micrometeorites. These are mainly protons (or hydrogen nuclei), ejected in a steady stream from the Sun at an average speed of 500 km/sec. The Moon has no protection against these fast-moving particles. In principle its gravitational or magnetic fields could deflect them, but both are too weak to do so.

Proton energies, and therefore the damage they do on impact, are very small, but the large influx more than makes up for their small size. Collision and radiation damage are of a more subtle kind than the gouging-out of craters. Most of the protons penetrate only a few millimetres into the surface material, and the main changes they make to its structure are chemical; internal structures of crystals in lunar rocks are sometimes dislocated, chemical compounds are formed and gases can be generated. The visible effect of the solar wind on the lunar surface is two-fold: 'bleaching' and 'cementing'.

An interesting effect, still to be investigated, is the possible storage of some of the energy of protons in the crystals of rocks. Part of this stored energy might be reclaimable for useful work.

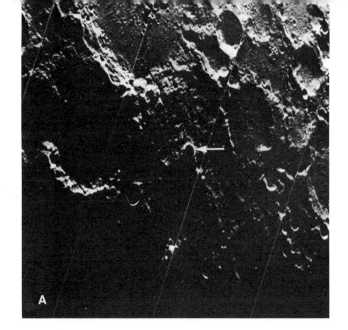

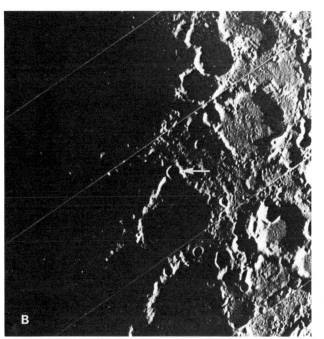

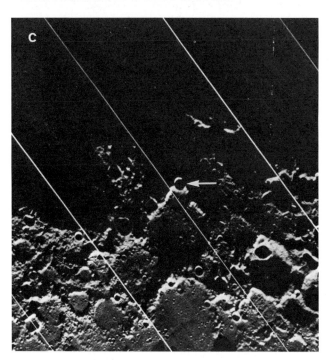

Plate 83 The Moon's north pole: the pole is located near the small circular crater at centre, and the region shown is about 360 km square. The direction of the Sun's rays changes, but at the pole the Sun is in all cases about 1° above the horizon. Many craters can be seen whose floors are never lit. Arrows show one such crater which always has its floor in shadow.

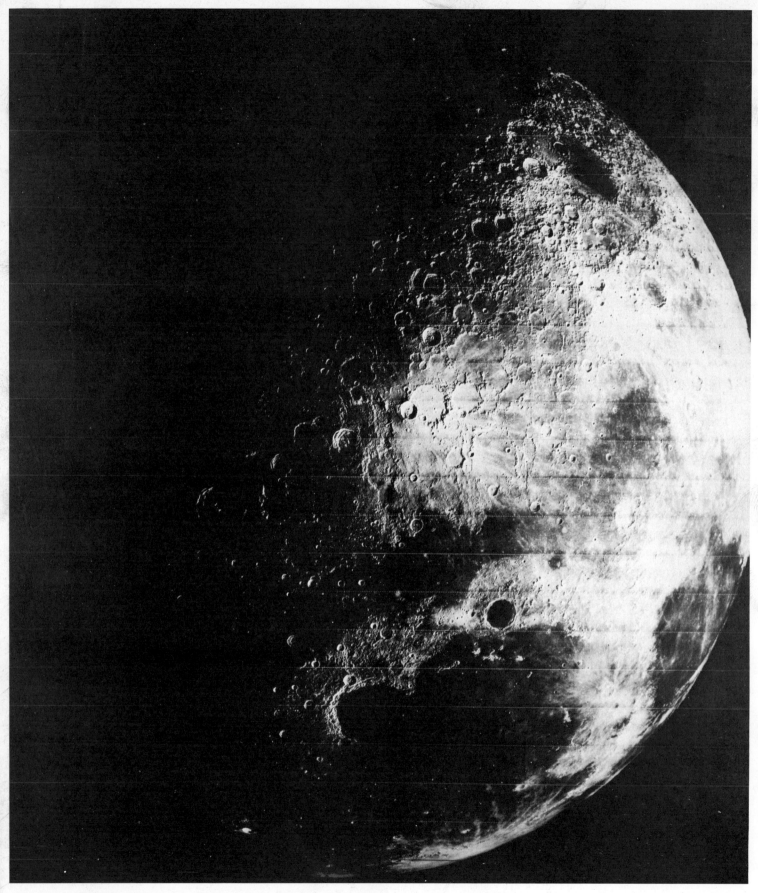

Plate 84 Orbiter view of Moon near the north pole. The dark mare at bottom right is Mare Imbrium; the bay on Imbrium's north-west side is Sinus Iridum. *NASA photograph.*

Plate 85 Ranger IX picture of Alphonsus, a crater 117 km across near the centre of the Moon's nearside disk. Plate 107 shows the detail of dark-halo craters indicated by the arrow. *NASA photograph.*

Plate 86 Detail of Plate 85 taken when the Ranger IX spacecraft was only a short distance above the lunar surface. *NASA Photograph.*

Plate 87 Lunar rock returned to Earth by Apollo 11 mission. Note the small craters on the rock's surface. *NASA photograph.*

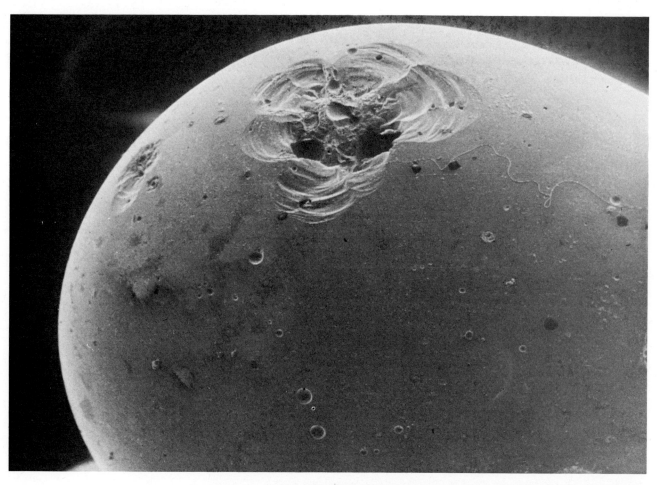

Plate 88 Electron-microscope photograph of the surface of a glass bead from the lunar regolith (Apollo 11). The crater at the top is a small impact crater less than 0.3 mm across. *Copyright Cambridge Electroscan.*

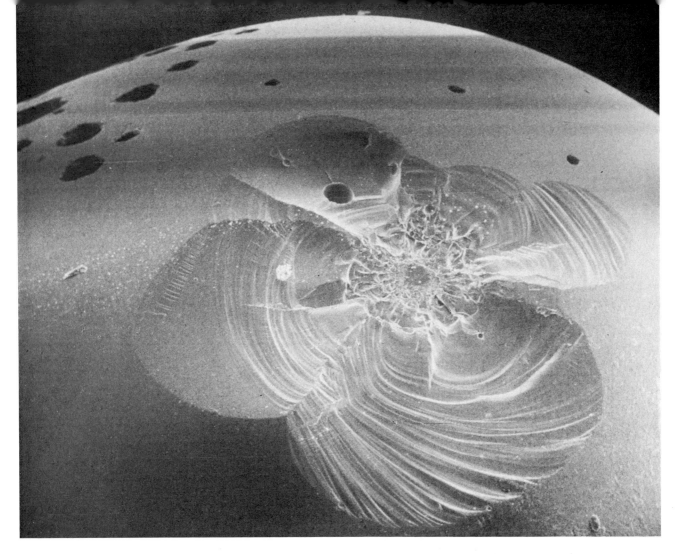

Plate 89 A close-up of small crater of same type as shown in Plate 88. *Copyright Cambridge Electroscan.*

Plate 90 The lunar Alpine valley. Oblique photograph looking south-west into Mare Imbrium. The mountains in the foreground are the Alps. Note the contrast between them and the younger dark mare plains. The Alpine valley is 15 km across. *NASA Orbiter V photograph.*

The Earth and Its Satellite

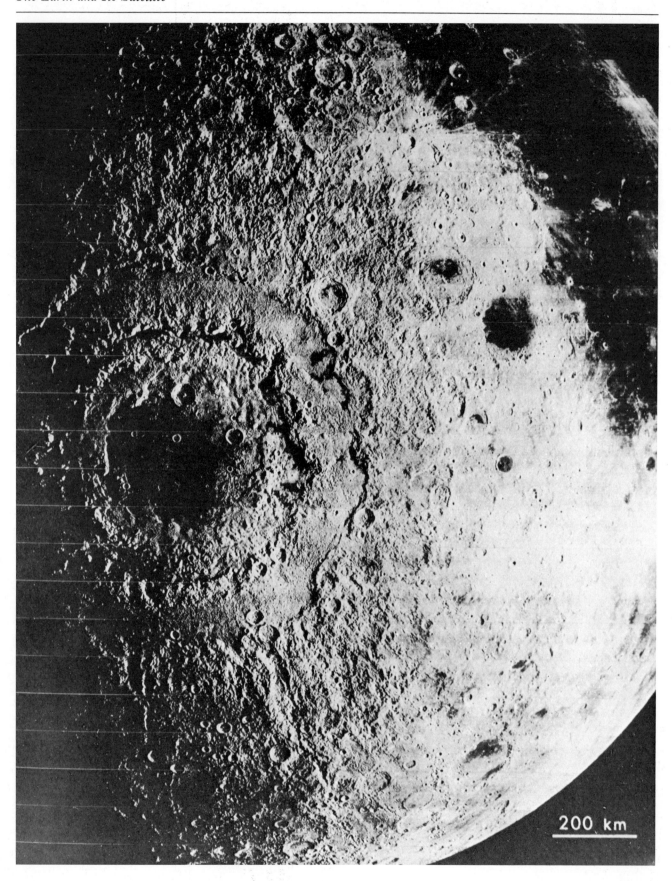

Surface of the Moon

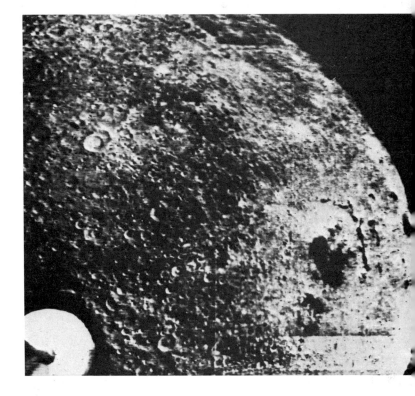

Plate 91 (above) Apollo 8 photograph of the east limb of the Moon. The circular mare at top left is Mare Crisium, that at centre is Mare Smithii. *NASA photograph.*

Plate 92 (opposite page) This splendid Orbiter IV photograph shows Mare Orientale, some 600 km across, on the west limb of the Moon. Unlike maria shown above, Orientale is only partly filled with dark marial material. *NASA photograph.*

Plate 93 (top right) Detail of part of area shown in Plate 92 taken with the high-resolution camera of Orbiter IV. The flow-like features are thought to be layers of ejecta thrown out during the formation of Mare Orientale by a gigantic impact. *NASA photograph.*

Plate 94 (right) Russian Zond photograph of the lunar farside, taken July 1965. *Novosti photograph.*

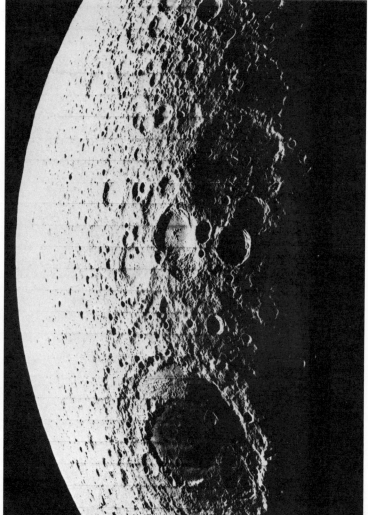

Plate 95 (above) Oblique photograph of the lunar farside taken by the Apollo astronauts. The large crater is Tsiolkovsky which is about 200 km in diameter and has a dark floor of lava. The crater itself was probably formed by a large impact. *NASA photograph.*

Plate 96 (left) Part of the lunar farside showing Mare Muscoviense (bottom). Note the densely cratered highland terrain of the farside. North is at the top. *NASA Orbiter V photograph.*

Plate 97 (above right) A large lava flow in Mare Imbrium. Note the lobate flow front which is about 30 metres high. The photograph represents an area of about 80 × 60 km.

Plate 98 (right) Numerous sinuous rilles near Prinz, the large crater (50 km across). Most sinuous rilles start in small craters. The rilles are thought to be collapsed lava tubes. *NASA Orbiter IV photograph.*

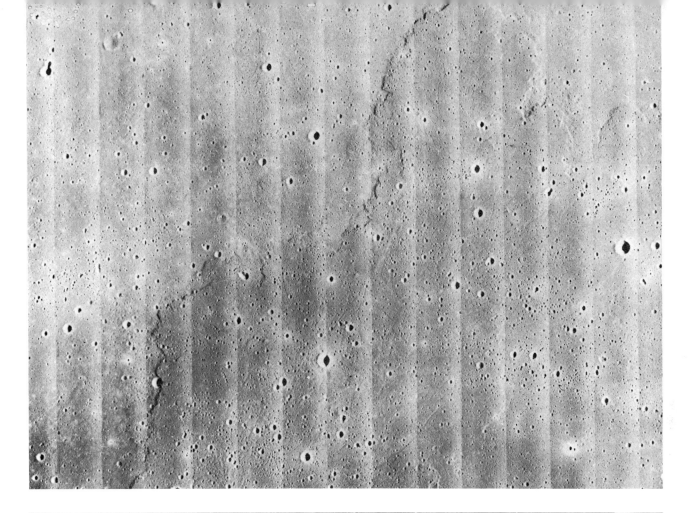

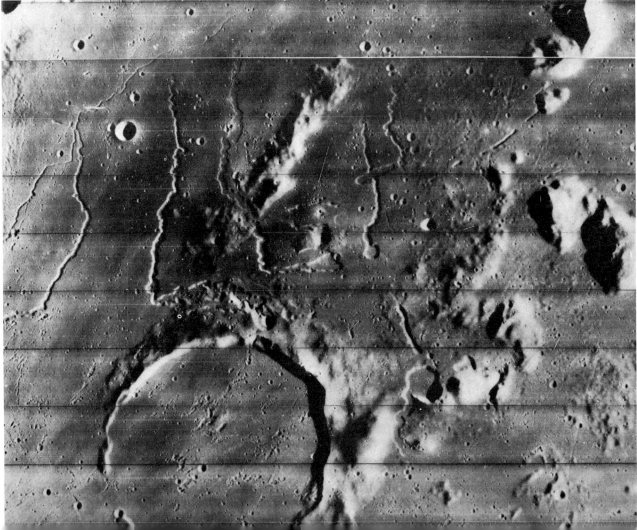

Chapter Nine

Geology of the Moon

J. E. GUEST

From the very first observations of the Moon's surface using a telescope it was obvious that the dominant features were craters. Thus, whereas the most active process that sculptures the Earth's surface is the erosion of rocks by running water, on the Moon the main process on all scales from the minute to the gigantic is one that produces craters.

It was quite natural for early lunar workers to think that all the lunar craters were formed by volcanism as this was the only crater-forming process known to them; at that time it was not known that the Earth and Moon were subjected to bombardment by METEORITES of all sizes, from dust up to large bodies several kilometres across. Before the rôle of meteorites was recognised it had been suggested that the lunar craters might have been formed by impacting bodies, but the idea was rejected because *it would be difficult to imagine whence those bodies should have come.*

Near the end of the nineteenth century G. K. Gilbert, the American geologist, promoted the idea that the craters were produced by the impacting of meteorites, and stressed the differences between many lunar craters and terrestrial volcanic craters. Since the beginning of this century the volcanic and impact theories for lunar craters have developed hand-in-hand—the volcanic theory being more readily accepted in Europe, the impact theory in North America. As one might expect, it is now clear that both processes have played important parts in shaping the lunar surface; but that, although volcanism has been of considerable importance, it has probably not produced the majority of lunar craters, most of which appear to be of impact origin.

As progressively more detailed pictures have been returned from the Moon during the past decade it has become apparent that cratering is much more important than was thought, even from the highest-resolution telescopic observations. Telescopes had shown craters down to about $\frac{1}{2}$ km in diameter; close-up photographs taken by Rangers 7, 8 and 9 showed that the surface was peppered with small craters down to a few metres across (*Plate* 86); soft-landing Surveyor craft showed small craters of less than a few centimetres in diameter. This very small-scale cratering was confirmed by rocks brought back by the astronauts (*Plate* 87). Even more surprising was the fact that small glass beads, themselves less than 1 mm across, also showed cratered surfaces, some of the craters being only a few thousandths of a millimetre in diameter (*Plates* 88 and 89).

The Maria and Highlands

Two types of very different terrain are distinguishable on the Moon (*Plate* 81): the dark areas, or MARIA, which are relatively level and have few large craters on them, and the brighter, densely cratered areas or HIGHLANDS. The maria are of two types: the circular ones like Imbrium and Crisium (see *Plates* 84 and 91), and the irregular depressions such as Oceanus Procellarum.

The maria and highlands not only differ in character but broadly represent two different periods in the history of the Moon. The highlands are for the most part old, and became heavily pockmarked by large craters early in the Moon's history. Impacts into the Moon by large asteroidal-sized bodies are thought to have formed the large

Plate XLI
Meteorites.
(A) Part of a stony meteorite which fell at Barwell, England on 24 December 1965. The meteorite broke into pieces shortly before it struck the ground. Note the dark fusion crust on the lower left of the specimen; this was formed by intense heat as the meteorite passed through the Earth's atmosphere.
(B) An iron meteorite that fell at Rowton, England, on 20 April 1876. The surface above the sixpence has been artificially cut, and etched with acid to show the characteristic criss-cross pattern of iron-nickel alloys. *Photographs by P. J. Elgar (courtesy of the British Museum).*

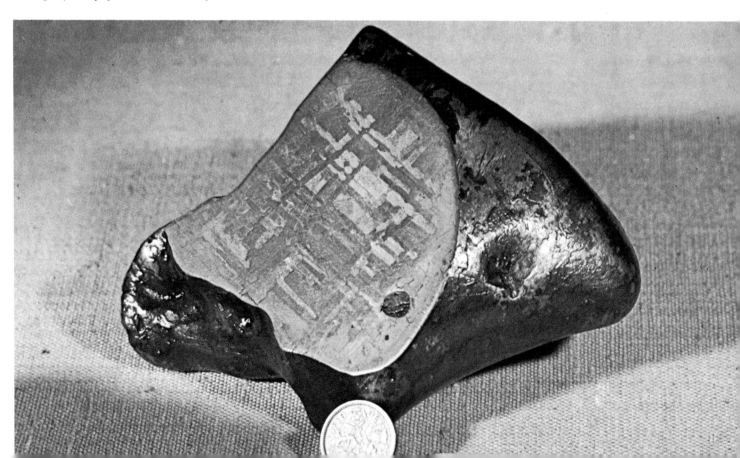

Plate XLII
A yellow octahedral diamond (80 carats) set in basaltic kimberlite from the Dutiotspan Mine, Kimberley, South Africa. *Photograph: J. W. Harris (courtesy of the De Beers Consolidated Mines Ltd).*

Plate XLIII
A metamorphic rock seen through a microscope. The rock has been cut in a thin slice so that light can pass through it. This rock crystallised deep in a mountain chain. The round grain in the middle of the picture is garnet, round it swing flakes of mica; these cause the rock to split like slate. *Photograph: R. Mason.*

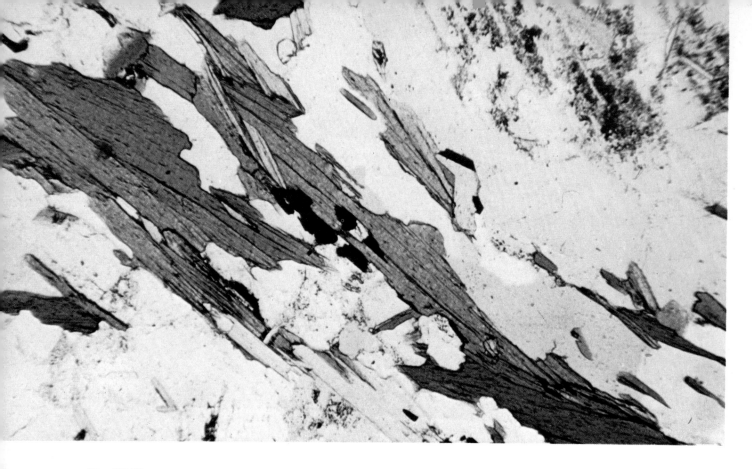

Plate XLIV
Thin sections of acid igneous rocks.
(A) Granite with two types of mica; brown biotite and colourless muscovite. The dominant components of this rock are feldspar and quartz.
(B) A glassy rock of similar composition to granite. The rock is mainly composed of glass rather than of crystals because it cooled rapidly as a lava before crystals could form. The one crystal of quartz in the section was corroded by the glass when the lava was hot. *Photographs: M. K. Wells.*

Plate XLV
Thin sections of basic igneous rocks.
(A) Basalt composed mainly of brightly coloured pyroxene and parallel-sided crystals of plagioclase with a black and white striped appearance.
(B) Gabbro, a coarsely crystalline form of basalt, probably formed by basalt cooling deep below the surface in a magma chamber. *Photographs: M. K. Wells.*

Plate XLVI
Thin section of lunar igneous rock collected from Mare Tranquillitatis by the Apollo 11 astronauts. Crystals of plagioclase (dark and white striped, long crystals) and pyroxene (brown) are clearly seen. Compare this rock with those in Plate XLV.
Photograph: S. O. Agrell, University of Cambridge.

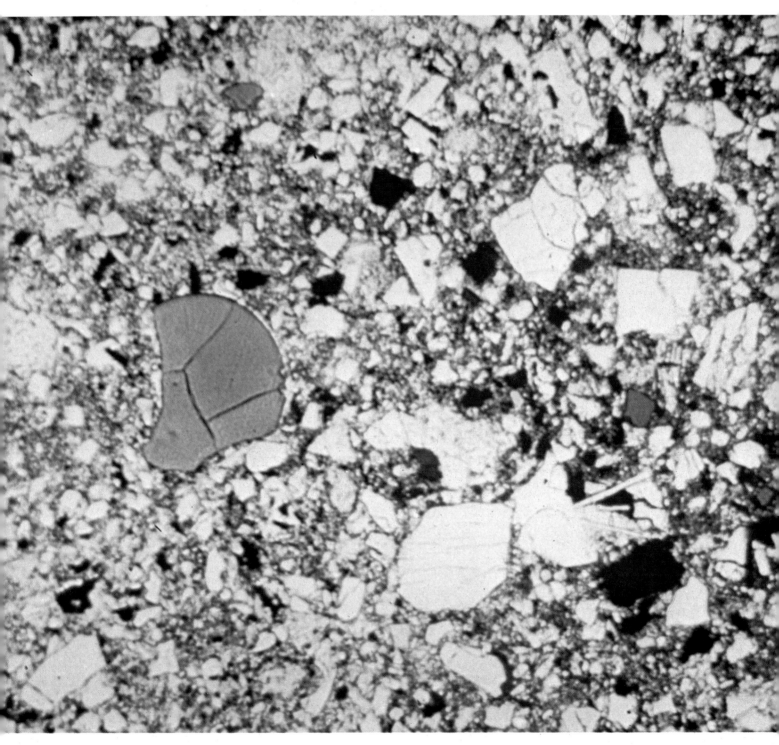

Plate XLVII
Lunar breccia from Mare Tranquillitatis. This rock consists of fragments of igneous rock, glass and spherical glass beads. Break-up and melting of igneous rocks by meteoritic impact is considered to have formed rocks of this type on the Moon.
Photograph: S. O. Agrell, University of Cambridge.

Plate XLVIII
Fine material from the lunar regolith.
(A) Lunar dust.
(B) Enlarged view of a small glass bead found in the fine dust above. These beads are thought to have been formed from small blebs of molten rock produced during a meteoritic impact.

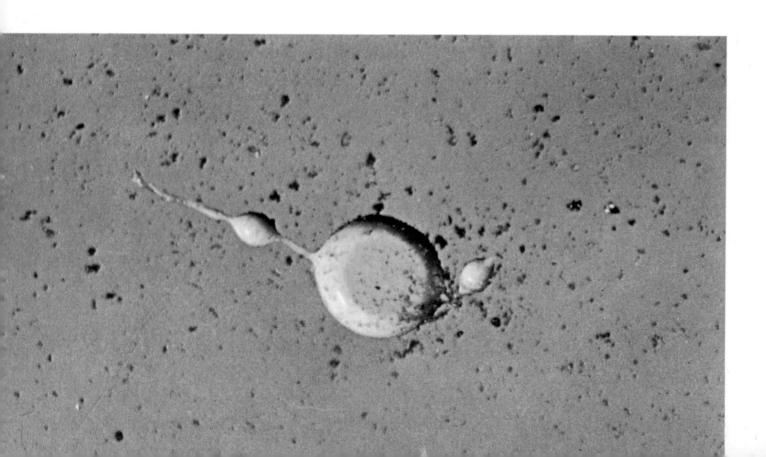

circular MARIAL BASINS, excavating those enormous craters and spreading great volumes of ejecta for hundreds of kilometres round them (*Plates* 92, 93). These catastrophic events that, to a large extent, gave the Moon its present geography occurred at an early stage in the life of our Earth-Moon system, probably more than 3,700 million years ago. Those marial basins occurring on the nearside were then explained? It appears that a small amount of dark material was formed in Orientale early in its history, but did not continue to be formed. In Imbrium dark material was produced over a long period of time and continued to form long after Orientale had been excavated. One must draw the conclusion from this that dark mare material results from a protracted process such as volcanism, rather than

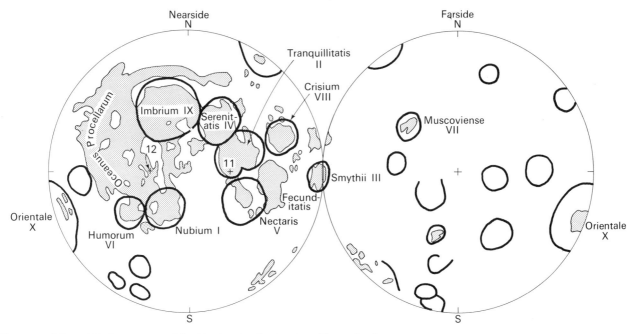

Figure 25   Maps of the lunar near and farside showing the location of large circular basins. Relative ages of the basins are given in roman numerals with Nubium as oldest (I) and Orientale the youngest (X). Dark mare-material is stippled.

filled with extensive, dark lava flows that we call the maria.

The large, circular marial basins were not all formed at the same time. Figure 25 gives the possible order of their formation. Some scientists have considered that the dark rocks of the maria were produced by the melting of rock by the heat generated when the circular basins were excavated by large impacts. If this were so then the age of individual maria would be the same as the basins with which they are associated. However, present studies suggest that not only are the dark marial areas of different ages but they do not correspond in age with the basins they fill. For example, the Orientale basin is younger than the Imbrium basin, and both these basins are younger than any of the other large basins on the Moon; but the dark mare surface in Orientale is much older than the equivalent material in Imbrium. How can this paradox be

from an almost instantaneous process like impact melting.

The Far Side

So far we have considered only the nearside of the Moon. The far side was totally unknown until Russian spacecraft relayed pictures from behind the Moon (*Plate* 94). The big surprise from these pictures was that unlike the nearside, the far side is composed almost entirely of heavily cratered highland terrain. Later, American probes confirmed this, showing that only a few of the larger craters contain dark material comparable to that of the nearside maria (*Plates* 95 and 96). The reason for this lack of mare material in far-side circular basins is unknown at the present time and stands as one of the fundamental problems of lunar geology.

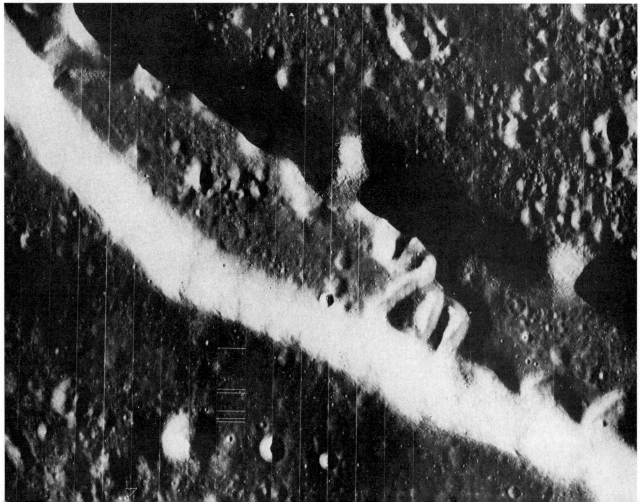

Geology of the Moon

Plate 99 (opposite) Schröters valley, a sinuous rille near Aristarchus (in bottom right corner). At its widest this rille is 5 km across. *NASA Orbiter V photograph.*

Plate 100 (below left) Close-up of Plate 99 taken with high-resolution camera. *NASA Orbiter V photograph.*

Plate 101 (right) Oblique Orbiter III photograph across Mare Tranquillitatis. The crater in foreground appears to have been cut in half by a large fault. Note the well-marked mare ridges. The distance on the horizon is about 280 km. *NASA photograph.*

Plate 102 (below) Orbiter V photograph of the volcanic complex of the Marius Hills in Oceanus Procellarum. Note the mare ridge complex running across the area. The sinuous rilles (or valleys) may be old lava tubes. Volcanic domes occur in a number of places. *NASA photograph.*

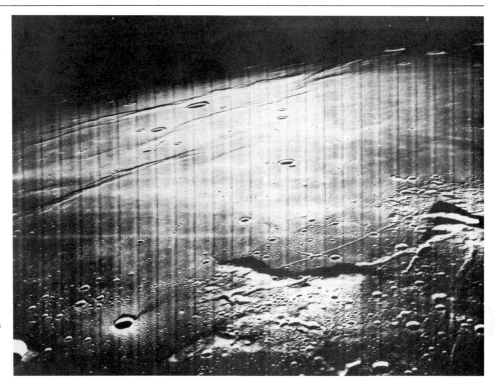

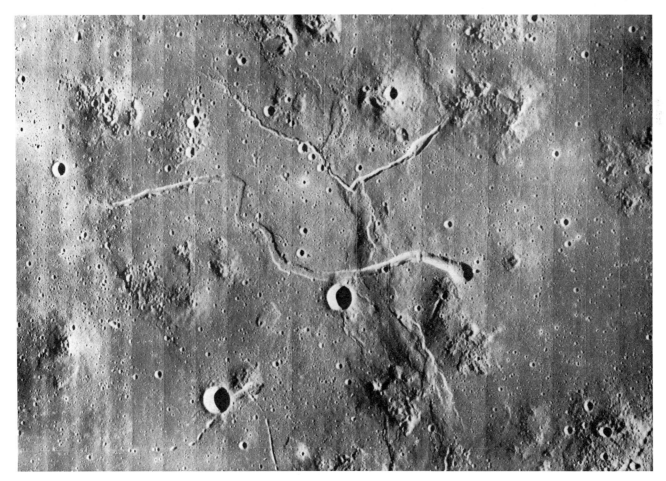

The Earth and Its Satellite

Plate 103 Vesicular lava from Mare Tranquillitatis. Collected during the Apollo 11 mission. *NASA photograph.*

Geology of the Moon

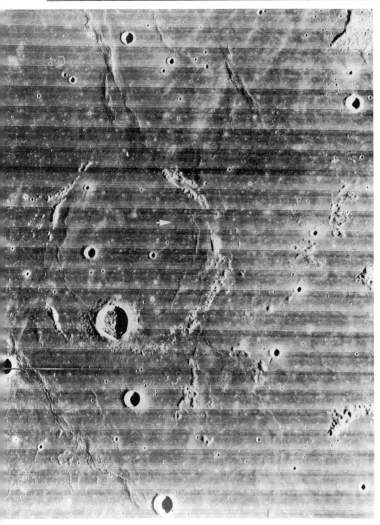

Plate 104   The Flamsteed P ring, a crater-like structure in Oceanus Procellarum. It has a diameter of about 100 km and was the site of the Surveyor 1 landing (arrow). North at top. *NASA Orbiter IV photograph.*

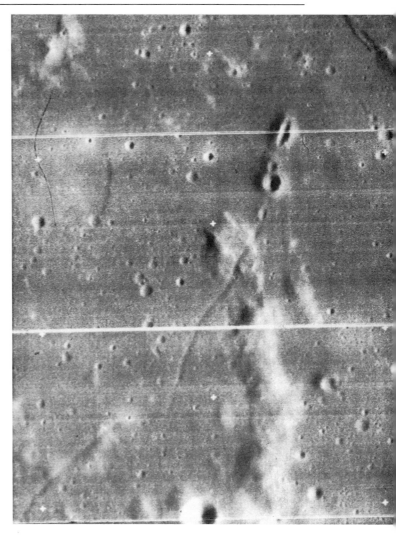

Plate 106   A fissure volcano near Flamsteed. Several cones have formed on a wide fissure. *NASA Orbiter IV photograph.*

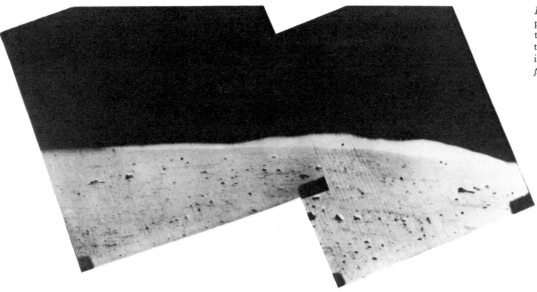

Plate 105   Television picture taken by Surveyor 1 to show the hills of part of the Flamsteed P ring shown in Plate 104. *NASA photograph.*

The Earth and Its Satellite

Plate 107 Dark halo craters in Alphonsus. These lie on graben structures. The dark material is thought to be volcanic ash thrown out from the crater. *NASA photograph.*

Geology of the Moon

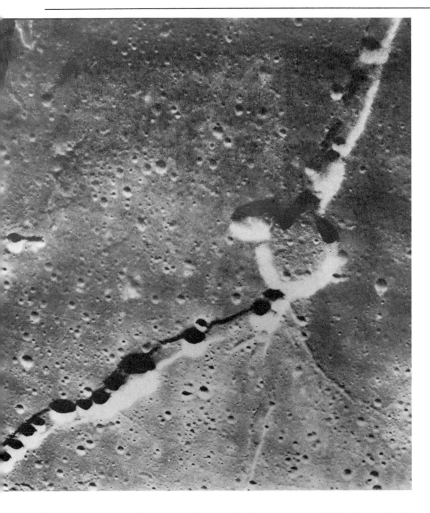

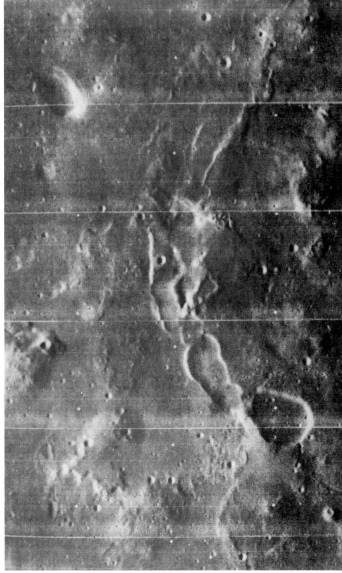

Plate 108 (*top left*) Hyginus rille. A graben structure near the centre of the Moon; thought to be formed by subsidences of troughs between parallel faults. The craters are clearly associated with the rille and were formed mainly by collapse. The largest crater is 8 km across. *NASA Orbiter IV photograph.*

Plate 109 (*above*) Caldera-like depressions near the Marius Hills in Oceanus Procellarum. These have diameters of about 8 km. Compare with the terrestrial caldera in Plate 20. *NASA Orbiter IV photograph.*

Plate 110 (*left*) Lunar domes in Mare Orientale. These resemble terrestrial shield volcanoes. *NASA Orbiter IV photograph.*

The Earth and Its Satellite

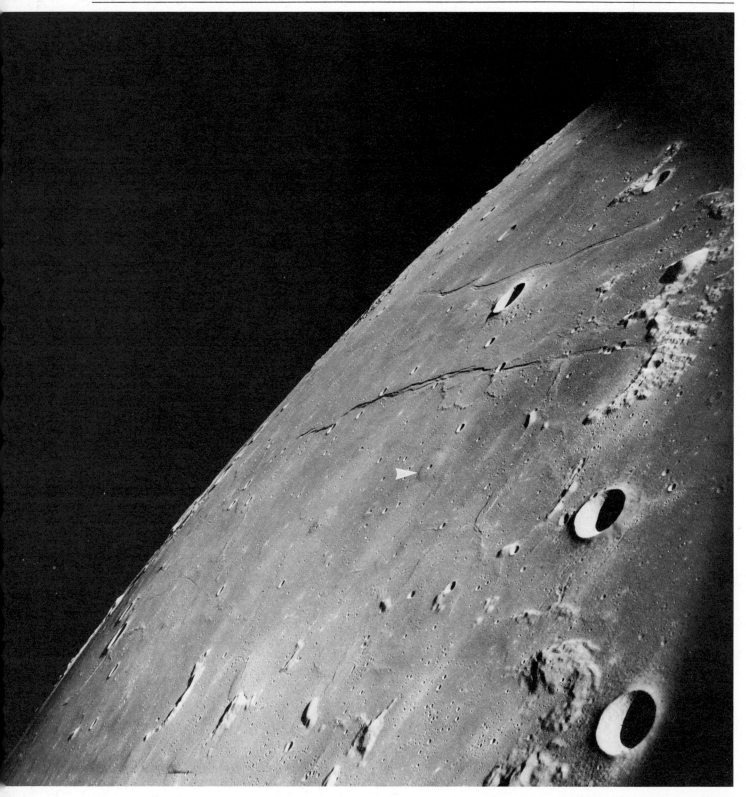

Plate 111   The Cauchy domes in western Mare Tranquillitatis. Note the central crater on dome marked with arrow, and the two graben fault structures. The dome marked with arrow is about 9 km across. *NASA Apollo 8 picture.*

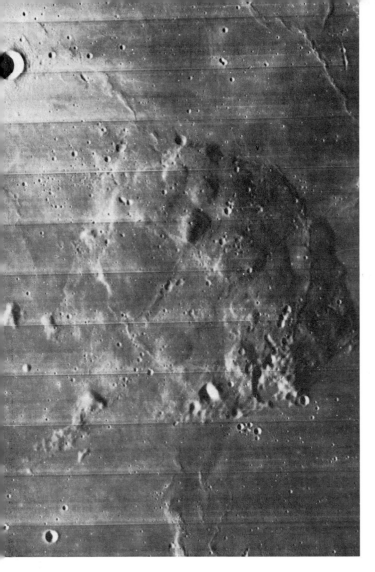

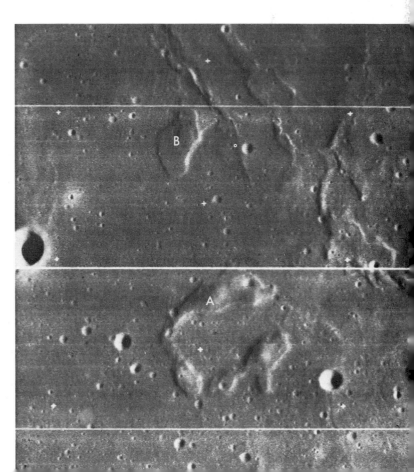

Plate 112 (*above*)   Rümker, a volcanic plateau some 40 km across in Oceanus Procellarum. Many low domes occur on the plateau. *NASA Orbiter IV photograph.*

Plate 113 (*top right*)   A volcanic cone (C) from which a thick lava has flowed. Note the mare ridge (R); this is probably an extrusion of lava along a fissure. It appears to have flowed into an older crater at (F). *NASA Orbiter V photograph.*

Plate 114 (*right*)   An extrusive ring of lava (A) and an apparent lava flow (B). The lava ring is a few kilometers across. *NASA Orbiter IV photograph.*

Lava Flows in the Maria

Following from the above arguments for a volcanic origin for material of the marial regions, it is necessary to look at these regions in detail to see if volcanism can be further demonstrated. The most conclusive evidence was provided by the samples brought back from Apollo missions. The first two manned landings were both in marial regions, one in Mare Tranquillitatis and the other in western Oceanus Procellarum. It was natural that mare sites should be chosen for the first landings because, although covered with craters on the small scale, they are much less rugged than highland areas. Rocks from these sites showed the maria to be underlain by lava flows of 'basaltic' composition.

Detailed study of Orbiter pictures had predicted that rocks of this type would be found. One of the best pieces of evidence discovered was the presence of large lava flows in Mare Imbrium. These had been seen from telescopic observations, but the return of higher-resolution photographs showed a series of extensive flows many tens of kilometres long and covering hundreds of square kilometres. The best of these is shown in Plate 97; it has a well-marked, lobate flow front which, from the length of the shadows it casts, is probably up to 30 metres high. This flow has all the characteristics of basaltic lava flows found making up extensive lava fields on Earth (*Figure* 15).

The flows in Mare Imbrium are large by terrestrial standards, suggesting that when they erupted the lava was able to flow rapidly over great distances, and was therefore more fluid than normal basaltic lavas on Earth. It was thus not surprising to find, from experimental work on simulated lavas of the same composition as Apollo 11 rocks, that lunar lavas could be much more fluid than those on Earth. These experiments showed that molten lava in the maria might have had the consistency of thick engine-oil, allowing it to spread even more freely over the lunar surface than the Imbrium flows did. Because of the low viscosity, well marked FLOW FRONTS would not have formed readily, possibly explaining why it is not possible to distinguish individual flows on many marial surfaces.

One of the puzzles about which there has been much controversy is the origin of the SINUOUS RILLES. These resemble meandering valleys, leading some workers to the conclusion that they were formed by rivers of flowing water (*Plates* 98, 99 and 100). This appears to be unlikely from our present knowledge of the Moon, and there is some doubt as to whether they have the shape expected for a river-cut valley on the lunar surface. One important observation is that they all start at their upper ends in well-defined craters—an unlikely feature for a river unless it had been formed by the melting of a layer of ice by volcanic activity below the surface. The general shape of these sinuous rilles is like that of lava tubes; these are formed by lava rivers flowing below a consolidated crust of lava, and if the crust collapses into the channel a sinuous valley remains. This analogy was suggested many years ago, but was not entirely acceptable because the lunar rilles are so much bigger than those of terrestrial lavas: while terrestrial lava channels or tubes may be up to several metres across, the lunar rilles are as much as a kilometre or more across. With a very fluid lava, however, such large rivers of lava might be expected, and with the low gravity on the Moon a large roof could be supported over the tube while the lava was flowing. Although this is a likely explanation for these features it does not explain all of them completely. They will remain a research topic for some time to come. However, it is possible that these sinuous rilles mark the positions of the many lava flows whose boundaries we cannot see.

Mare Ridges

The majority of the large lavas cannot be traced back to a source crater or volcano. We are thus left with the problem of how they were erupted. The most likely sources for these lavas are the long mare ridges that are ubiquitous features in the maria. They are usually several hundred metres wide and some tens of metres high. They rarely form one continuous ridge, but are made up of many shorter lengths aligned *en echelon* or in parallel patterns to give long ridge complexes. Many ridges are thought to be formed by extrusions of lava from fissures. This would have been a more viscous lava than that which formed the extensive flows. In some places the lower margins of the ridges are flows which can be seen to fill small craters in the underlying surface. Other ridges, especially those which are just broad swells, may be zones where lava has not reached the surface but has pushed up the surface over the fissure; often ridges of this type have smaller ridges on top where lava appears to have just broken through to the surface. It is presumed that after the eruption of the large lavas from fissures, these fissures were filled with more viscous MAGMA to give the present ridges.

Rocks from the Maria

Two types of rock were returned from the Apollo 11 and 12 missions: one was a fragmental rock which will be described later, and the other was volcanic lava and thought to represent the bedrock of these areas. The lavas consist of crystals of PLAGIOCLASE, PYROXENE and ILMENITE, together with smaller quantities of OLIVINE and less common minerals. The principal minerals of the rocks are all found in terrestrial igneous rocks (see Chapter Five). Only a few minerals unknown on Earth have been found. One of these has been named *Armalcolite* after the three astronauts of Apollo 11: *Arm*strong, *Al*drin and *Col*lins.

The textures in the rocks (i.e. the size of the crystals and the way they are arranged) suggest that they crystallized as very fluid lavas—in agreement with the laboratory experiments mentioned previously—and that they cooled fairly rapidly. Melting experiments suggest that as flowing lavas they had a temperature of about 1,200 °C to 1,300 °C. One characteristic of the crystalline rocks is the presence of vesicles (*Plate* 103). Such vesicles are typical of lavas on Earth, where they are formed by gases 'boiling' out of the lava.

Although the rocks from Apollo 11 collected in Mare Tranquillitatis are all of similar chemical composition it is clear that they cannot be representative of all maria: first because remote-controlled analyses by Surveyor craft have shown that although Tranquillitatis has a notably high titanium content other maria have not; and secondly because Apollo 11 rocks are clearly potential sources for rocks of other compositions. Between the main minerals of the rock there is a filling of material differing in composition from the bulk of the rock; while the bulk is 'basaltic' the material in the gaps is essentially 'granitic'. If this 'granitic' fraction had been separated off from the rest of the rock it could well have been erupted as a completely different type of lava, quite separately from the rest of the 'basaltic' lava. Since we know that processes of this type occur on Earth it appears likely that it has occurred on the Moon. It was, therefore, not surprising to find a much wider variation in composition from rocks brought back from eastern Oceanus Procellarum by the Apollo 12 astronauts. The silica content of these rocks ranges from less than 40% to about 61%. Thus we can see that lunar rocks so far collected cover more than half the terrestrial range (see Chapter Five). Further collections may show a much wider range. As we shall see in the next section this is of considerable importance in interpreting the various volcanic landforms seen on the Moon.

TABLE 4

	Apollo 11	Apollo 12	
	Typical basaltic lava (weight %)	Typical basaltic lava (weight %)	Intermediate rock (weight %)
SiO_2	45·0	40·0	61·0
Al_2O_3	9·0	11·2	12·0
TiO_2	10·0	3·7	1·2
FeO	17·0	21·3	10·0
MgO	8·0	11·7	6·0
CaO	9·5	10·7	6·3
Na_2O	0·6	0·45	0·7
K_2O	0·2	0·07	2·0
MnO	0·4	0·3	0·1
Others	0·8	0·6	0·5
Total	100·5	100·02	99·8

Chemical analyses of rocks collected from the lunar maria. Note the high TiO_2 content of the Apollo 11 rock; and the wide differences between two of the rocks from Apollo 12 (especially SiO_2, FeO and K_2O).

As well as the 'basaltic' rocks there were also small fragments of a rock known as *anorthosite*: this rock consists almost entirely of the mineral plagioclase and has a density of 2·9. Its composition corresponds well with the only analysis we have of a highland region, that carried out by Surveyor 7 in the vicinity of the crater Tycho. It has been suggested that the highlands, therefore, have an anorthositic composition. Later Apollo missions, or possibly unmanned Russian missions, may provide us with the answer to the problem of the composition of the highlands; it may well be much more complex than the simple anorthositic model so far suggested.

Volcanic Craters and Vents in the Maria

With so much volcanic material in the maria we may expect to find volcanic craters. This is not so

Plate 115 (left) A small, fresh, bright ray crater a few tens of metres across. This is considered to be an impact crater. The other craters on this picture are thought to be older impact craters that have become denuded. *NASA Orbiter photograph.*

Plate 116 (below left) A marial surface peppered with small craters of different sizes. Close examination of the picture will show different shapes of crater ranging from sharp fresh ones to subdued old ones. The area shown is about 4 × 6 km. *NASA Orbiter II photograph.*

Plate 117 (below) Close-up of a small crater in Oceanus Procellarum. It has a diameter of about 150 metres; and its terraced shape is probably due to a layer of denser rock below the surface. Craters of this type are common on the Moon and can be used to determine the depth to the underlying bedrock. *NASA photograph.*

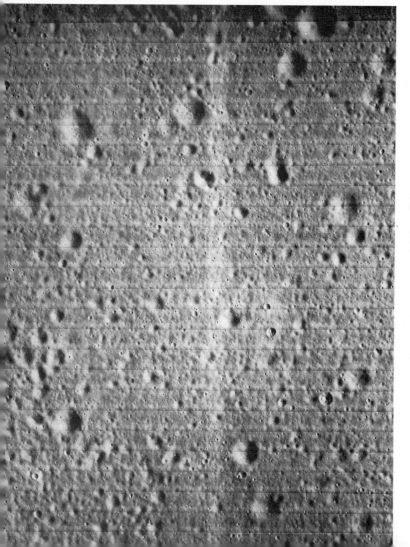

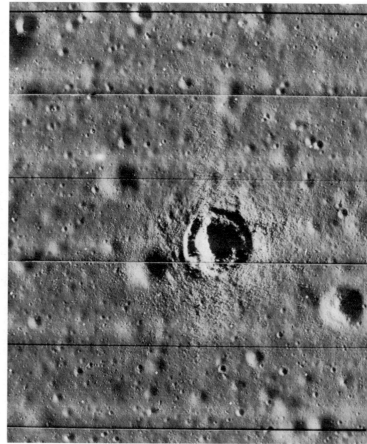

Plate 118 This Ranger photograph was taken just 5.5 seconds before the craft crashed into the lunar surface. Smallest craters shown are about 10 metres across. *NASA photograph.*

Plate 119 Surveyor 1's footpad depressed in the fine-grained material of the lunar regolith. Grains smaller than 1 mm diameter can be resolved near the footpad. *NASA photograph.*

easy as it may appear because the whole surface of the Moon is peppered with craters, many of which are probably of impact origin. This overprint of impact craters tends to hide those that are volcanic. However, some types of crater are so different from those produced by impact that they are readily identifiable.

Volcanic craters formed on large fissures or faults are the most easy to recognise. Fissure volcanoes occur in a number of maria. Essentially these consist of a line of cones joined by a fissure (*Plate* 106). *Dark halo* craters are also often recognisable as being volcanic by their location on GRABEN structures (straight rilles). In Alphonsus there are a number of these (*Plate* 107). They owe their name to the fact that they are surrounded by a dark ejecta blanket, unlike the majority of small craters which, when fresh, have bright ejecta. Similar to these dark halo craters are the rille craters (*Plate* 108). They also occur on graben, and some are probably of the same type as the dark-halo craters, but older, the dark material having been mixed into the REGOLITH. Others have almost no ejecta around them and are probably entirely the result of collapse. Most of the craters associated with graben tend to be irregular in shape in contrast to the highly circular impact craters.

With the eruption of large volumes of lava into the maria, portions of the lunar surface may be expected to collapse into the emptied MAGMA CHAMBERS. Large collapse craters of this type are known on Earth as *calderas* (see Chapter Five). Features similar to calderas of the type found in terrestrial basalt terrains are relatively common on the Moon (*Plate* 109). These are not unlike the calderas of Hawaii.

Calderas are not the only features of the maria that have direct counterparts in basaltic terrains on Earth. One characteristic feature of the maria is the presence of several types of domes. Many of these are low, with shallow slopes of just a few degrees (*Plates* 110, 111). They are often up to 10 km across and a few hundred metres high. In some cases there is a summit crater. These are identical in appearance to volcanoes known as *shield volcanoes* on Earth, made up of numerous flows making up a pile of volcanic material over a vent-complex.

Other domes are steep-sided, apparently being made up of more viscous lava squeezed out to form what is essentially a large lava flow, too viscous to flow very far and so piled up as a dome over the vent from which it was extruded (*Plate* 102). Sometimes small, thick flows occur; these originate from well-defined cones of pyroclastic material thrown out by explosive activity during the formation of the lava flow (*Plate* 113). More viscous lavas were also apparently erupted from ring fractures to give circular ridges of lava (*Plate* 114); the ring factures probably formed over a magma chamber.

From this brief review of types of volcanoes in the maria we can see that volcanic activity has taken a variety of forms, ranging from extensive floods of fluid lava to the eruption of viscous lava as small flows. Explosive volcanism has also produced several types of crater. In Chapter Five we saw that the type of volcanic activity depends on the conditions of the lava at the time of eruption: for example its viscosity, temperature and gas content. Composition of the lava plays an important role, different physical conditions being to some extent related to composition. The wide variety of the types of volcanism in the maria thus implies that lavas of a wide range of composition may be expected; an inference supported by the rocks so far brought back from the maria.

Small Cratering and the Regolith

It is obvious from higher-resolution photographs of the Moon that the surface is peppered by numerous small craters (*Plate* 118). These are nearly always circular. The fresh ones are usually bowl-shaped with sharp outlines, and have bright ejecta blankets extending into fine radial rays (*Plate* 115). The majority of these are thought to have been formed by meteorites hitting the Moon's surface. With time they suffer bombardment from the continuing impacts of small meteorites, and as they become older, the sharp features are worn off and the bright ejecta churned up. This process continues until what was a fresh, sharp crater becomes a subdued wreck of its original form. Many regions thus have a complete range of craters from old to very young.

Some craters are not bowl-shaped, but have concentric ridges or terraces inside them (*Plate* 117); this characteristic probably results from the craters having been formed in layered rocks, some layers being more resistant than others.

Bombardment by meteoritic material (see Chapter Eight) also breaks up the rocks on the surface and produces a layer of fragmental material covering the whole Moon. The thickness of this layer

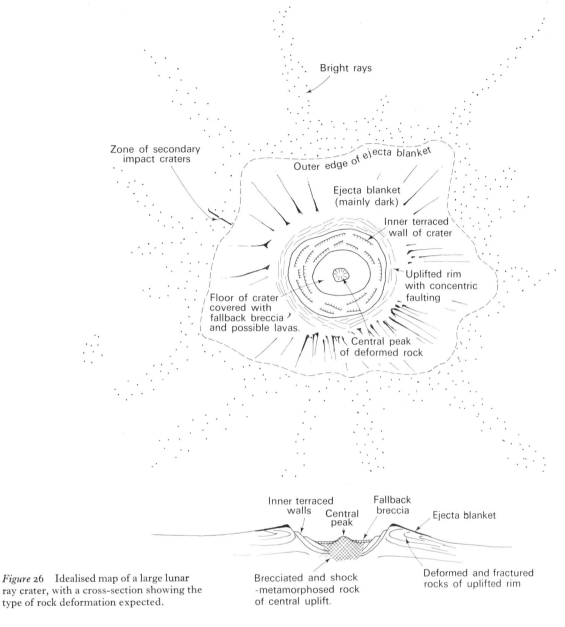

Figure 26   Idealised map of a large lunar ray crater, with a cross-section showing the type of rock deformation expected.

will to some extent be a measure of the length of time the rock layer has been subjected to attack. Areas with high crater density are older and have a thicker fragmental layer. Careful counts of numbers of craters and measurements of the thickness of the fragmental layer can provide a useful indication of the relative ages of different surfaces.

The fragmental surface layer is called the REGOLITH. Although its existence had been predicted previously, the television pictures from Surveyor soft-landers first demonstrated its properties. It was seen to consist of rock fragments of various sizes set in a poorly consolidated, fine-grained powder (*Plates* 119, 120), into which the footpads of the craft sank up to a few centimetres. All the samples brought back from Apollo missions were collected from the regolith. From these samples it was shown that not only were there blocks of crystalline lava described previously, but there were less resistant blocks of BRECCIA, a fragmental rock made up of particles of different sizes, welded together. These breccias have a similar composition to the lavas, and appear to be made up entirely of fragmented, and re-melted lava that is now in the form of glass (*Plate* XLVII). The matrix of the breccias is the powdery material of the regolith, except that in the breccias the powder and larger fragments are welded together to give a coherent rock. The observation that much of the regolith is derived by break-up of the underlying bedrock of lavas conforms to the observations earlier that meteoritic bombardment has churned up the whole lunar surface.

It is now known that when meteoritic impact

Plate 120 A narrow trench dug in the regolith by Surveyor 3's mechanical digger. *NASA photograph.*

Plate 121 A lunar rock seen by the camera of Surveyor 1. The length of the rock is about 45 cm. *NASA photograph.*

Geology of the Moon

Plate 122 View from Surveyor 1. The large blocks mark the line of a large ancient crater several hundred metres in diameter. Far rim of crater is on horizon. *NASA photograph.*

Plate 123 'Tranquillity Base', the first manned landing site. One astronaut is removing equipment from the lunar module. The instrument in the foreground is a camera for photographing the lunar surface. Note the astronauts' footsteps in the soft surface of the regolith. The surface is strewn with half-buried pebble-size fragments. Note the large blocks to the right. *NASA photograph.*

takes place the high temperatures and pressures developed at the site of impact alter the impacted rocks, changing the structure of the minerals and, in the extreme, melting them: this process is known as SHOCK METAMORPHISM. The products of shock metamorphism can be demonstrated at known impact craters on Earth such as Meteor Crater, Arizona (*Plate* 64) and also in nuclear craters (*Plate* 65). Both the fines and the breccias in the lunar rocks contain shock-metamorphosed material—further evidence that the regolith was formed by meteoritic churning.

The most extreme form of shock in the regolith results in the production of glass; this occurs as fragments and small glass beads (*Plates* XLVII, XLVIII). All the glass appears to have been formed by melting of the lavas: specks of iron have been added to the beads, presumably representing material from some of the meteorites.

As mentioned earlier in this chapter, the rock fragments in the regolith have small craters on their exposed surfaces. These are thought to have been excavated by the impact of micrometeorites. Again these will have an erosional effect on the small scale. Rocks that originally had angular surfaces are worn down until they have a rounded appearance. This phenomenon was inferred from Surveyor observations which showed that sharp, fresh craters had more angular blocks than the older soft craters.

When regolith is formed on sloping surfaces, such as the inner walls of craters and on sides of mare ridges, it slides slowly down the slope giving the surface a characteristic ridged pattern (*Plate* 125) similar to that produced by soil-creep on Earth. Much of the creep of the regolith on the Moon is probably brought about by shaking of the ground by moonquakes and nearby meteorite impacts. Plates 126 and 129 show the even more dramatic rolling of boulders down slopes.

Large Ray Craters

Anyone observing the full Moon, even with binoculars, cannot help noticing the three main large ray craters: Aristarchus, Copernicus and Tycho. They are characterised by widespread systems of bright rays extending radially for hundreds of kilometres (*Plate* 81). All three of the above craters are young relative to the majority of lunar features; this is demonstrated by the observation that the rays overlie most other major features. Inside the rays there is a darker zone consisting of fragmental material surrounding the crater (*Figure* 26). This is known as the ejecta blanket and mantles craters and other topographic features in the underlying terrain, subduing them, finally burying those near to the crater where the thickness of the ejecta blanket exceeds the extent of the topographic relief below it. Near to the crater lip the ejecta blanket has a hummocky top surface, but it levels out away from the crater except where it overlies other craters.

The closest man has ever got to one of these craters is through the eye of the television camera on Surveyor 7, which landed on the ejecta blanket of Tycho (*Plates* 130, 131). The ejecta blanket was shown to consist of fragmental material like the regolith, but more blocky and much thicker. An example of this is shown in Plate 131, which also includes a large block thought to have produced the impact crater near to it. Because the block appears to have bounced out of the crater instead of burying itself as would have happened if it had hit the surface at very high speed, it is considered to have been thrown out from Tycho when this large crater formed. A glance at Plate 127, a telescopic picture of Copernicus, shows that near the outer edge of the ejecta blanket (at about one-crater diameter) there are numerous small craters forming an annulus of high crater density. It is generally thought that these craters were all formed in the same way as the small one seen by Surveyor near Tycho, although many would have been produced by ejected blocks that had much higher velocities: they are termed *secondary impact craters*.

All the features so far described suggest that large ray craters were formed by one single, gigantic explosion: the ejected blocks falling to form smaller secondary craters all round the main crater, and the other ejecta falling to give the ejecta blanket. The rays were formed by long streamers of fine material thrown considerable distances away from the explosion. Although we have never witnessed such large explosions on Earth, we have a partial analogy in nuclear explosions, which also produce craters with all the features described here (*Plate* 65). What caused the lunar explosions? Some suggest that it could have been volcanic activity, but although this cannot be neglected entirely, our present knowledge of the solar system suggests the most likely cause to be the impact of a meteoroid (or perhaps a cometary nucleus, see COMET). Such a body colliding with the Moon would produce a colossal explosion of the magnitude necessary to excavate craters of 100 km or more in diameter.

We know from a study of both nuclear craters and old impact craters on Earth that when great explosions of this type take place the rocks below and around the crater become highly folded and later fractured by the force of the explosion. The type of folding is shown in Figure 26. Fracturing of the predicted type is seen quite clearly in the lunar craters; there are close-spaced concentric fractures round the lips of the craters and also fine radial fractures.

When large explosions take place, be they volcanic, nuclear or impact, a *base-surge* cloud forms. This is made up of solid fragments suspended in gas, and flows away from the crater like a liquid, slowly depositing the fragments on the ground below it (similar to *nuées ardentes*, Plate XIX). The layer of material so formed resembles a flow in many ways. Some extensive flow-like units around craters such as Aristarchus and Tycho may be of this origin (*Plate* 132). Other flows were clearly formed by viscous liquids: these are most probably younger lava flows, or flows of molten rock squeezed out at the time of formation of the crater. Semi-molten rock also spread over the floors of the craters (*Plate* 134). This again may be volcanic or molten rock produced by the heat of the impact.

Craters in the Highlands

Many different types and ages of crater are found in the highlands. Some, like Tycho, are fresh craters formed after the marial basins had been filled with lava; others are more denuded and were formed before the maria were as we see them now. It is possible to show how craters change with age as they become eroded (*Plate* 137). The oldest craters, that have been deeply eroded, probably once looked like the ray craters and are most likely eroded impact craters (*Plate* 138). Others have been partially buried by later lava flows (*Plates* 137, 140).

There are other types of crater that still pose a problem (*Plate* 139). These usually have some of the characteristics of impact craters but contain also much evidence that extensive volcanism has occurred, although the type of volcanism is different from anything known on Earth. At present it is not certain whether these craters were formed by impact and have been later strongly modified by volcanic activity, or whether they are entirely volcanic.

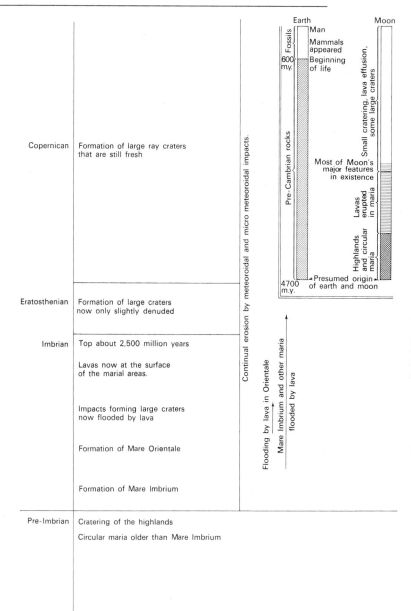

Figure 27 The lunar time scale. This is compared with the Earth's time scale in the insert.

Whatever conclusion one comes to, the evidence so far available suggests that the highlands are generally very old; certainly older than 3,700 million years if we accept the age derived from Apollo 11 rocks. Many conclude from this that the older parts of the highlands represent the primordial surface of the Moon. If this is so it may be that the Earth, too, once had a cratered surface long before the appearance of life.

Plate 124 One of the Apollo 12 astronauts inspecting the Surveyor 3 craft. This had been on the lunar surface since April 1967. In the background is the lunar module that the astronauts had travelled in. *NASA photograph*.

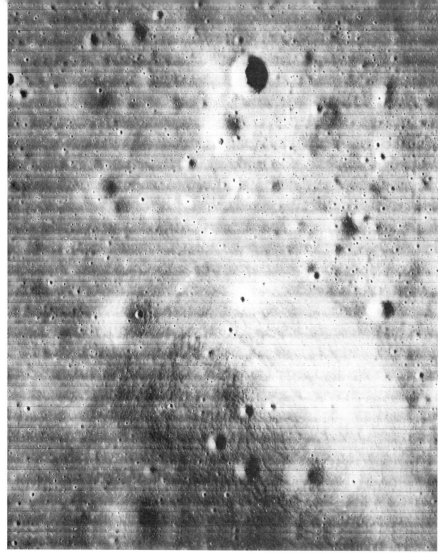

Plate 125 A hill, 2½ km across and a few hundred meters high in the south-eastern portion of Mare Tranquillitatis. The hill is covered by a layer of regolith which has moved downhill to give the fine ridge pattern on the hillside. Downhill creep of the regolith probably results from small moonquakes and local meteorite impacts. *NASA Orbiter photograph.*

Plate 126 (*below*) Tracks of two boulders that have rolled down a slope on the lunar surface. The boulders can be seen at the end of the tracks. *NASA Orbiter photograph.*

Plate 127 A telescopic view of the large ray crater Copernicus. This crater is about 90 km across. Near the crater the ejecta blanket has a hummocky appearance. At a distance equivalent to one crater diameter there are numerous small secondary craters. Outside these bright rays can be seen radiating from Copernicus.

Plate 128 An Orbiter 11 oblique view into Copernicus. The hills in the middle of the floor of the crater are clearly seen. Rising behind them is the inner terraced wall of Copernicus; and behind that, the Carpathian Mountains forming the southern boundary of Mare Imbrium. The hummocky terrain in the foreground is the ejecta blanket outside Copernicus on its southern rim. The cross on the far inner wall marks the area of Plate 129. *NASA photograph.*

Plate 129 Part of the inner wall of Copernicus. The sunlit wall at the top of the picture is that marked with a cross on Plate 128. Boulders, some as large as large apartment buildings, have rolled down the slope and are seen on the level ground at the foot of the slope. *NASA photograph.*

Plate 130 The large ray crater Tycho seen from an Orbiter craft. This crater has a diameter of over 40 km. The cross on the northern rim marks the landing site of Surveyor 7—see Plate 131. *NASA photograph.*

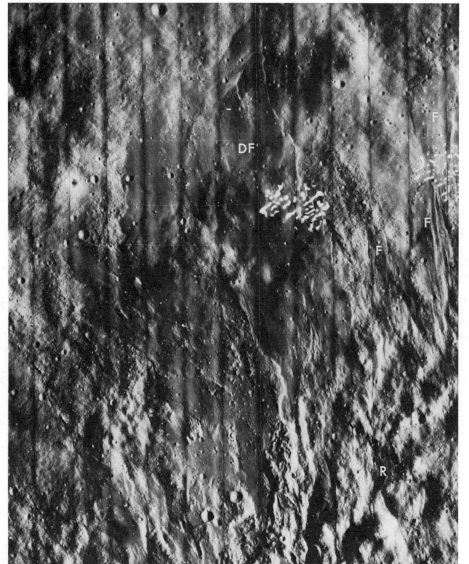

Plate 131   Blocks strewn on the ejecta blanket surface near the crater Tycho. These blocks are thought to have been thrown out of Tycho during the gigantic impact explosion that produced the crater. This photograph was taken by Surveyor 7 which landed at the point marked with a cross in Plate 130. *NASA photograph.*

Plate 132   Part of the northern rim of Aristarchus (see Plate 75C). In this area the hummocky part of the ejecta blanket (R) extends into thin flows (F) thought to be of fragmental material and to have formed from base surge clouds. The flows are only about 1 metre thick. Dark flows (DF) may have a different origin. *NASA Orbiter V photograph.*

Plate 133 Thick flows on the rim of Tycho. These were formed from viscous fluids. Some scientists think they are lava flows erupted after Tycho has formed, but others think they are mudflows. *NASA Orbiter V photograph.*

Plate 134 Part of the floor of Tycho's crater. It appears from the flow patterns that this was formed from a molten rock; but it is not clear whether this was lava or rock melted by the impact that formed Tycho. Hills at the foot of the main crater wall are seen at top left. *NASA Orbiter V photograph.*

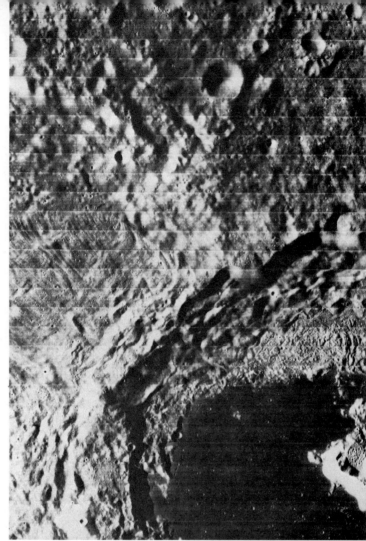
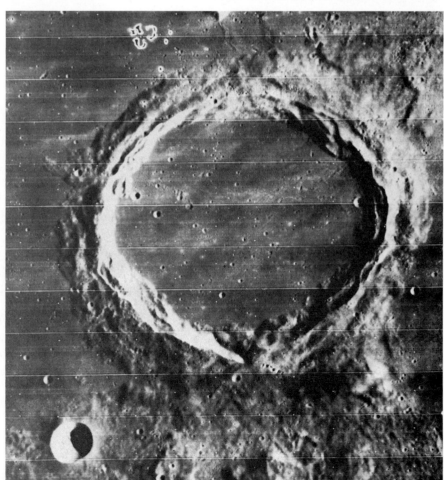

Plate 135 Tsiolkovsky, a lunar farside crater with a diameter of about 200 km (see Plate 95). This crater, like Copernicus, Aristarchus and Tycho, is thought to have formed by impact. It was later flooded by dark lava flows in the floor of the crater. Note that Tsiolkovsky has cut a much older denuded crater to the west. This old crater is now known as Fermi. *NASA Orbiter photograph.*

Plate 136 (*above right*) A close-up of Tsiolkovsky taken at the same time as Plate 135. The inner terraced walls of the crater are like those of Copernicus (Plate 127) and the material on the floor of the crater at the foot of the slope is similar to that of Tycho shown in Plate 134. However, dark lavas appear to have covered much of the original floor. The large flow-like feature outside the crater to the north-west is probably a large landslide. *NASA Orbiter III photograph.*

Plate 137 Archimedes, a large crater 81 km across in the south-eastern edge of Mare Imbrium. This crater also has all the features of a ray crater, but has been flooded by large areas of dark mare lavas covering much of the ejecta blanket and floor of the crater. *NASA Orbiter IV photograph.*

Geology of the Moon

Plate 138 An old denuded crater on the lunar highlands. *NASA photograph.*

Plate 139   Gassendi, a large crater some 110 km across on the northern edge of Mare Humorum. The origin of the crater is unknown: some think it is an impact structure, others that it is volcanic. There appears little doubt that, whatever the origin of the crater, it was later flooded by volcanic material. The huge cracks resembling terrestrial graben remain another problem of lunar geology. *NASA photograph.*

Plate 140   A photograph by the Apollo 8 astronauts of the crater Goclenius which is 52 km wide. This crater, like some of the others in the photograph, has been partly submerged by dark lava flows forming the level mare plains. Later tectonic activity cut the area with long rilles or graben. *NASA photograph.*

Chapter Ten

Dating Rocks from Earth and Moon

M. R. WILSON

Give a man a rock and he will want to know its age. There are two ways of answering. One is to say that it is older or younger than another rock (perhaps because the rock is from a lava flow which obviously overlies older material) and thus give the rock a position on a scale of relative ages. Such a scale has been worked out in fine detail for Earth rocks, particularly for those younger rocks which contain fossils. High-quality photographs of the Moon's surface can be used to distinguish the relative ages of craters and other features.

The alternative answer is to give the rock an age in terms of years, or more usually, in millions of years (see *Figure* 28).

To determine Earth and planetary ages in terms of years has long been an aim of research workers. The attempted ways have been described as 'hour-glass methods', being based on the observation of a process which continues at a known rate. Early hour-glass methods included the measuring of the rate at which rivers add salt to the oceans; the rate at which the Earth was thought to be cooling; and the rate of accumulation of sediments in the ocean basins. These methods were doomed to failure because the process rates were variable and subject to too many disturbances.

The discovery of radioactivity in 1896 led to the development of the modern hour-glass methods, based on radioactive decay. Certain elements such as uranium are present in forms which are fundamentally unstable, decaying to form other elements. If the rate of decay is known and the ratio of 'parent' to 'daughter' element can be measured, then in some cases the age of the system can be determined. First suggested by Rutherford in 1905, these methods have made great advances in precision and accuracy in the past 20 years, and, being applicable to a wide range of materials over the whole of geological time, have tremendously aided and stimulated geological research.

What is Meant by Radioactive Decay

All substances are made up of millions of ATOMS of one or more ELEMENTS. For some elements, such as gold, all the atoms have the same MASS, but for many elements the atoms have a range of masses. Atoms of a single element which differ in mass are termed *isotopes*. For example, rubidium is represented by two isotopes with masses 87 and 85. By analysing an element on a MASS SPECTROMETER the relative abundances of the different isotopes can be measured.

Unstable isotopes decay to atoms of simpler structure. This rearrangement is accompanied by a release of energy (radiation). The original unstable atom is termed the 'parent' and the new atom the 'daughter'.

What is the Rate of Radioactive Decay?

Each unstable isotope has its own characteristic rate of decay which is simply expressed as its half-life. The half-life is the time required for a given number of atoms of an isotope to decay to half that quantity. Suppose there is an isotope with a half-life of one year. If we commence with 16 atoms of parent (and a complete absence of daughter) then after one year there will be 8 atoms of parent left and 8 atoms of daughter. After a further year the 8 atoms of

The Earth and Its Satellite

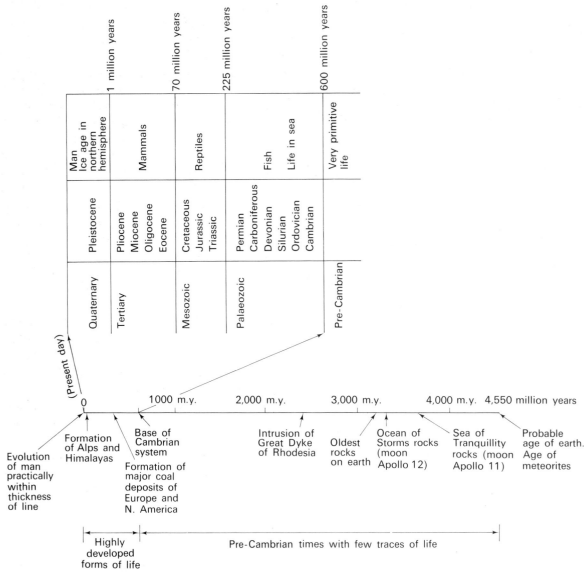

Figure 28 Ages of Earth and Moon rocks.

parent will have decayed to 4 atoms and there will be 12 atoms of daughter (*Figure* 29).

The beauty of radioactive decay for the geologist is that the decay rates are not affected by any known external physical or chemical factor. The rate of decay on the ocean floor, at the centre of the Earth and in outer space are all identical for a given isotope.

Decay Schemes Useful to the Geologist

Of the large number of unstable isotopes only a few are suitable for the geologist. Some isotopes decay too quickly, some too slowly, while in some cases the parent is extremely rare and in others the daughter isotope is already present in the rock when the latter crystallised. Some decay schemes are useless because it is physically impossible to measure the rate of decay with sufficient accuracy. The useful decay schemes are listed in Table 5, and these all fulfil the above criteria.

How to Date a Rock

Determining the age of rock is basically a matter of measuring the ratio of parent-to-daughter isotope and applying the appropriate formula. But what if there is some daughter isotope present initially?

TABLE 5

The main methods of radiometric age determination (m.y. = million years)

Parent	Daughter	Half-life	Application
Potassium^{40}	Argon^{40}	1,300 m.y.	Biotite, muscovite, hornblende, nepheline (minerals), basalt (rock)
Rubidium^{87}	Strontium^{87}	47,000 m.y.	Muscovite, biotite, feldspar, whole rock analyses of granites and other igneous rocks and minerals
Uranium^{238}	Lead^{206}	4,510 m.y.	Zircon, sphene, pitchblende
Uranium^{235}	Lead^{207}	713 m.y.	

What if the system has been disturbed so that some of the parent or daughter has been removed? Just what does a date mean when it has been determined?

To answer these questions a practical case of potassium-argon dating is examined. Suppose magma is generated at depth. It rises and is emplaced near the surface where it solidifies *quickly*. All the time potassium^{40} is decaying to argon^{40}. While the magma is liquid, the argon, being a gas, escapes. Even when the magma has crystallised the argon continues to escape because the hot crystals are unable to retain the gas. Eventually the crystals cool to a temperature at which the outward diffusion of the argon is blocked. This temperature, termed the *blocking temperature*, is about 500 °C for the mineral HORNBLENDE and as low as 200 °C for BIOTITE. Once this blocking temperature is reached radiogenic argon begins to accumulate. Thus the age calculated from the parent-daughter ratio (using the precisely determined half-life) is the time elapsed since the mineral cooled sufficiently to allow the radiogenic argon to accumulate. While the rock is above this temperature all radiogenic argon produced is lost to the atmosphere. Evidence for this is given by the observed fact that where an intrusive body has cooled slowly over several millions of years then the measured 'age' for minerals such as hornblende with a high blocking temperature is greater than that for minerals with low blocking temperatures such as biotite, as illustrated in Figure 30.

Providing the rock is kept cool (less than about 100 °C) and is not subject to weathering on the Earth's surface none of the radiogenic argon is likely to escape from such minerals as biotite, MUSCOVITE or hornblende. If these conditions are fulfilled and no argon escapes from the rock, it is said to have behaved as a 'closed system' with respect to argon.

It must always be understood that potassium-argon dates give the times elapsed since cooling, not since crystallisation. Only when the time taken to cool is less than the analytical error will the two times be essentially the same. Thus a date of 50 million years plus or minus one million years on a rock which cooled in a matter of 10,000 years can be taken as the date of crystallisation.

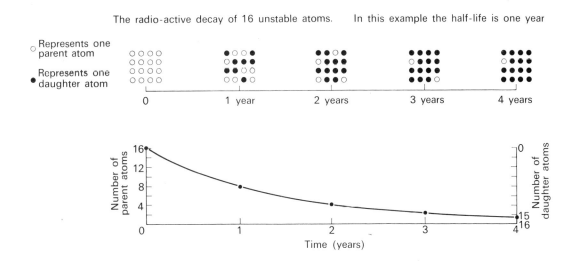

Figure 29 The rate of radioactive decay.

The Earth and Its Satellite

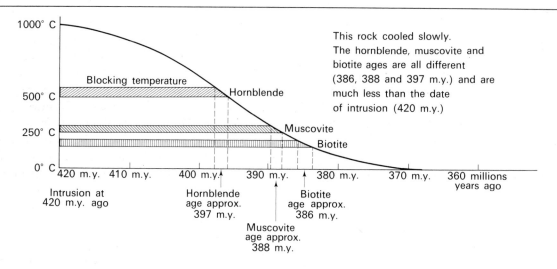

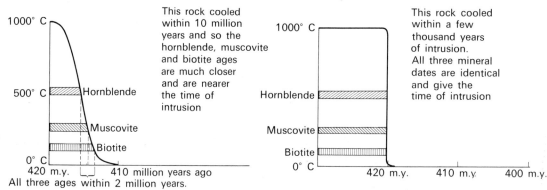

Figure 30   On these diagrams the cooling of a rock body is illustrated. The vertical axis shows the temperature of the rock and the horizontal axis indicates time (before present day). The horizontal lines indicate the blocking temperatures for the different minerals. Above this temperature all radiogenic argon escapes from the crystal. The potassium-argon date for a mineral gives the time elapsed since the rock passed through the blocking temperature.

What Rocks and Minerals can be Dated using the Potassium-argon Method?

Some rocks and minerals are unsuitable because the argon will escape from them at or near room temperature. These behave as 'open systems' with respect to argon. Potassium-feldspar is such a mineral, and unsuitable rocks include all glassy (uncrystallised) rocks, weathered rocks and those containing potassium-feldspar. Most other rocks and minerals are suitable if they contain over 0·1% potassium, but precise analysis is easier if they contain at least 1% potassium. Biotite, muscovite and hornblende are widely used minerals, while basalt is successfully dated as a whole rock.

The Rubidium-strontium Method

Rubidium^{87} decays to strontium^{87}. Unfortunately, the majority of rocks and minerals already contain strontium, including strontium^{87} at the time of crystallisation. Therefore it is obvious that it is impossible just to measure the ratio of rubidium^{87} to strontium^{87}. An allowance has to be made for any strontium^{87} present initially. This can be done easily if there is a mineral in the rock which contains no rubidium. The strontium^{87} in this mineral will not have increased in quantity with time, and the ratio of strontium^{87} to other strontium isotopes such as strontium^{86} will have remained constant (*Figure* 31). In rubidium-rich minerals radiogenic strontium⁸⁷ will be added to the initial strontium⁸⁷ and so the ratio of strontium⁸⁷ to strontium⁸⁶ will increase. By comparing this ratio with the strontium^{87}/strontium^{86} ratio from a rubidium-free mineral the actual *increase* in strontium⁸⁷ can be measured (*Figure* 31).

In practice several minerals from a rock are analysed and the determinations plotted on a graph (*Figure* 32) to form an ISOCHRON. On this graph the horizontal axis gives the ratio of rubidium to strontium while the vertical axis gives the strontium isotope composition. Any mineral can be represented on this graph by one point. Suppose that there are several minerals in a rock, all with different ratios of rubidium to strontium, and at the time of crystallisation they all have the same strontium isotope composition. If the analyses of these minerals at the time of crystallisation were plotted on the graph the points would all lie on a horizontal straight line. Consider the development of each mineral in time.

The mineral with no rubidium, plotting at the far left of the line, stays in the same position since the strontium isotopic composition remains identical. The mineral at the far right, with a high rubidium content, changes its strontium isotope ratio markedly through decay of the rubidium⁸⁷. Intermediate minerals change strontium isotope composition proportionately. At the end of a given period if the system has remained closed then each point (mineral) will lie on a new straight line. The slope of this line (the isochron) is proportional to the time elapsed since all the minerals had the same isotopic composition.

When an isochron is a plot of different minerals from one rock sample it is termed an 'internal isochron'. A second type of isochron is termed a 'whole rock isochron'. As a magma is intruded into the Earth's crust it may crystallise out as several slightly different types of rock (see Chapter Five). At the time of crystallisation these will all have

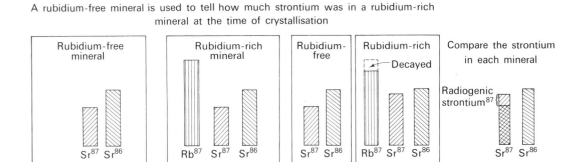

Figure 31 A rubidium-free mineral is used to tell how much strontium was in a rubidium-rich mineral at the time of crystallisation.

Figure 32 An isochron.

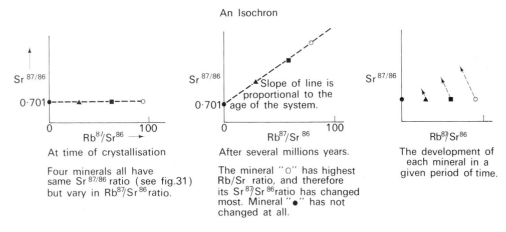

The Earth and Its Satellite

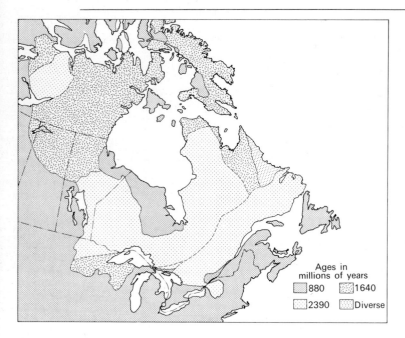

Figure 33 Ages of the last orogenic episode in each of the major structural provinces of the Canadian shield.

identical strontium isotope compositions, because isotopes tend to behave identically in all physical and chemical processes. The different fractions of rock may well have different strontium and rubidium contents. Each rock can be plotted as a point on the isochron graph to give a whole-rock isochron. Such an isochron can be determined only if it is known that all the rock samples are derived from one magma. Whole-rock isochrons give ages closer to the time of crystallisation of a magma than internal isochrons because specimens of whole-rock (say 1 kg in weight) become closed systems at much higher temperatures than minerals.

Uranium and Thorium-lead Methods

Here there are three independent schemes utilising the decay of thorium^{232} to lead^{208}, of uranium^{235} to lead^{207} and of uranium^{238} to lead^{206}. These are used on rare uranium- and thorium-rich minerals which can be found in granites among other rocks.

The Contribution of Geochronology to the Earth and Planetary Sciences

The full exploitation of the possibilities of age-dating has enabled not only igneous but also metamorphic and tectonic events to be timed. World-wide correlation is possible over the entire range of geological time. Sufficient determinations are now available to place the stratigraphic time-scale on a quantitative basis. Thus, for example, an intrusion dated at 160 million years and known to occur near the base of the Upper Jurassic (a unit of geological time defined by its fossil content) provides a fixed point on the time-scale. Of even more importance is our new ability to date and correlate the 3,000-plus million years of Pre-Cambrian history (*Figure* 33).

The student of the large-scale features of the Earth's crust is given valuable new data. The matching of ages on opposite sides of the Atlantic (*Figure* 34) is a fine piece of data which can be used in favour of continental drift, while some estimate of the rate of movement of the crustal plates is given by the increasing age of volcanic islands away from the mid-oceanic ridges (*Figure* 35). Geochronology has shed new light on orogenic belts (see Chapter Four) indicating the timing of events within them, the total duration of activity and the pattern of cooling after the activity.

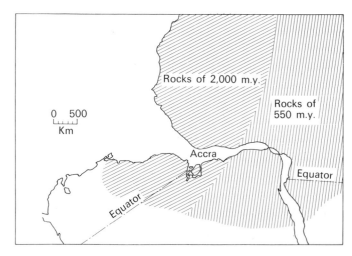

Figure 34 Map to show how rocks in South America and Africa agree in age when the continents are placed together in their pre-continental drift position.

To many people the most fascinating application of geochronology is the date of origin of the Earth, Moon and meteorites.

Age of the Earth

The best attempts to date the origin of the Earth have been by assuming that the parts of the solar system had a common origin and that the planets, asteroids and meteorites came from an homogeneous source. Because some meteorites contain no thorium or uranium the isotopic composition of the lead in them has stayed constant in time and is probably typical of the primordial leads in the solar system. On Earth the lead has changed its isotopic composition as a result of additions of radiogenic lead from decay of uranium and thorium. Because we know both the present isotopic composition of lead and the approximate concentrations of uranium and thorium on Earth we can estimate the time elapsed since the Earth had lead of the same isotopic composition as the primordial meteorite lead. Such a figure is 4,550 ± 50 million years.

This figure is much greater than the oldest rocks so far seen on Earth. Within most of the continents there are ancient cores with well-described rocks giving reliable dates of 2,400 million years. Within these cores are even older rocks, with ages up to 3,200 and 3,300 million years. It is hardly surprising that such rocks are not well preserved. Little is known about them and the interpretation of the radiometric dates is difficult.

Meteorite Ages

Meteorites have been studied by all three methods. The best determinations are rubidium-strontium internal isochrons. Most determinations give ages around 4,600 million years, plus or minus 100 million years.

Results from potassium-argon dating are not so reliable because there is evidence of loss of argon or potassium or both in many meteorites. While this 'open-system' behaviour is interesting from a chemical point of view it tells us little about the true age.

The Age of the Moon

The first two expeditions to the Moon brought back samples from the Mare Tranquillitatis and the Oceanus Procellarum. These have been analysed by several laboratories using different methods. As with meteorites the best determinations are rubidium-strontium internal isochrons. Five rocks from the Mare Tranquillitatis each have an age of around 3,650 million years. Two rocks from the Oceanus Procellarum have ages around 3,300 million years. These ages have analytical errors of about ± 100 million years.

It has been suggested that the lunar REGOLITH is a representative mixture of all the lunar crustal rocks,

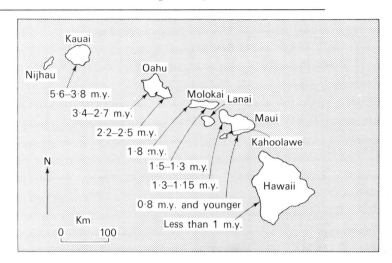

Figure 35 Ages of the Hawaiian volcanic islands. It appears from these dates that volcanic activity has migrated from the north-west to south-east.

thoroughly pulverised and mixed by countless meteorite impacts. The only way a rubidium-strontium age can be determined from such material is to assume a value for the isotopic composition of the strontium at the time of the Moon's formation. Certain types of meteorites (basaltic achondrites) have virtually no rubidium and so their strontium has remained more or less unaltered since the time of their formation. If meteorites and the Moon had a common origin then this value for primitive strontium can be used to calculate a 'model' age for the mixture of rock fragments which may indicate when Moon and meteorites parted company. Such a date is 4,500 to 4,600 million years. Since the same procedure can be applied to the very different systems uranium-lead and thorium-lead with similar results the basic assumption of common origin is given support.

Potassium argon determination on Moon rocks is hampered by the fact that the rocks have obviously lost radiogenic argon, probably as a result of heating and shock on meteorite impact. A method designed to eliminate this effect gives ages around 3,700 million years for the Mare Tranquillitatis rocks in good agreement with the rubidium-strontium results.

Uranium-lead analyses have not been too satisfactory since precise analysis of the very small samples is difficult.

It is now accepted that the Moon probably formed some 4,500 million years ago and that there was igneous activity for at least 300 million years

The Earth and Its Satellite

between 3,650 and 3,300 million years ago. Sampling of the lunar highlands could produce rocks even older than those already dated and give valuable insight into the Moon's early history.

Conclusion

The fascination of studying any subject in depth is that one is almost overwhelmed with facts, all with different degrees of validity and all coming from quite different fields of research. Most statements about the Earth and Moon rest on assumptions, some valid and some rather shaky. A long, revered hypothesis can be disproved by one rock sample and then what a rush of pens to paper! Weak arguments can be reinforced when several independent lines of evidence all point in the same direction. This is so with attempts to determine the age or origin of the Earth and Moon. The present accepted date of 4,550 million years is based on assuming that the Earth, Moon and meteorites had a common origin. But this assumption works independently for both rubidium-strontium and uranium-lead systems. The determination of internal isochrons (which do not depend on the above assumptions) for meteorites of up to 4,600 million years is a third support.

The more recent histories of the Earth and Moon are documented with more reliability. We know of rocks on Earth which are 3,200 million years old and have undergone highly complex evolution. (See Chapters Five and Six). Thus the Earth has always been a 'live' planet, recycling and reprocessing its material. Moon rocks have ages between 3,300 and 3,650 million years, and studies of the isotopic composition, chemical composition and behaviour under high pressure and temperature all show that the Moon rocks, while not so highly evolved as Earth rocks, are not particularly primitive. Although apparently dead for a long time the Moon was active once.

But every question answered raises ten new ones, and only new observations and more refined techniques coupled with man's ingenuity will permit progress.

Glossary

Words contained in the glossary are indicated in the text by SMALL CAPITALS.

Albedo A measurement of the reflecting power of an object; highly reflecting bodies are said to have high albedoes, and poorly reflecting bodies to have low ones. Some common substances have the following albedoes: black velvet 0·02; granite 0·25; chalk 0·4. Planetary albedoes are: Moon 0·12; Earth 0·38; Venus 0·85; Mars 0·14. Note the higher albedoes of planets with atmospheres.

Apogee The point in the orbit of a satellite (natural or artificial) when it is at its greatest distance from the Earth (see *Perigee*). For an orbiting lunar satellite the term is *apolune*.

Ash Small particles, usually less than 4 mm in diameter, erupted from a volcano by explosive activity.

Ash Flow The most dangerous of volcanic phenomena. It consists of a turbulent mixture of gas and ash, that flows like a liquid down the flanks of the volcano or along a gently sloping surface. Such flows deposit extensive flat sheets of ash; if the ash is still hot on deposition, the particles weld together to give a lava-like rock. Ash flow tuffs are usually called *Ignimbrites* in Britain.

Asteroid (Minor Planet) A small celestial body in the solar system normally having an orbit between those of Mars and Jupiter. The largest (Ceres) has a diameter of 700 km. Only 10% of the known asteroids exceed 30 km diameter.

Atmosphere The envelope of gas surrounding a planet. The Earth's atmosphere has the following composition: nitrogen 78%; oxygen 21%; argon 0·9%; carbon dioxide in small amounts and varying amounts of water vapour. The atmosphere of the Sun extends throughout the solar system. At the distance of the Earth it is extremely rarefied, containing only 10 particles per cubic centimetre.

Atom The smallest divisible unit retaining the characteristics of a specific element. Atoms are defined by their atomic number and mass. The chemical properties of an atom depend on the atomic number.

Barycentre The centre of mass of an orbiting system of bodies. The barycentre of the solar system is the point around which all the planets move. It is not fixed with respect to the Sun because it depends on the relative positions of all the bodies in the solar system at a given time.

Biotite A dark mica that is a hydrated silicate of iron, magnesium and potassium in varying proportions. It occurs as an original constituent of many types of igneous and metamorphic rocks (see *Mica*).

Breccia A rock composed of rock fragments (usually angular) cemented together to form a coherent rock. The fragments are derived from pre-existing rocks. The lunar breccias are made up of rock broken from the underlying bedrock (lavas, in the case of the maria) and shock-altered rock, often in the form of glass.

Caldera A large volcanic crater usually exceeding 5 km in diameter, formed by collapse of the surface into an underground cavity produced by rapid eruption of large quantities of magma, or withdrawal of magma from a magma chamber. Caldera subsidence may be accompanied by strong explosive activity in certain types of eruption.

Comet A member of the solar system consisting of a nucleus of small solid particles of frozen gases and dust. Near the Sun vaporisation takes place to give a rarefied gas 'tail' pointing away from the Sun. These tails are characteristic of comets and may have lengths of millions of kilometres. Some comets are visible to the naked eye.

Coriolis Forces An effect of the Earth's rotation, discovered by the French scientist C. G. Coriolis. An object moving north or south from the equator will maintain its initially greater peripheral velocity and will appear to curve either to the right or left (according to which hemisphere it is moving into) when observed from the surface of the Earth.

Density The ratio of the mass to the volume of a substance. This is expressed as kg/cubic metre. Densities are often quoted as *specific gravity*, the ratio of the density of a substance to that of water—thus water has a specific gravity of 1.

Dyke A near-vertical sheet of igneous rock. Dykes are usually only up to a few metres wide and are formed by the intrusion of magma into fissures; some reach the surface to give rise to an eruption of lava, normally as a fissure eruption.

Dynamo A device that generates an electric current.

Earthshine When the Moon is in the New Moon phase the Earth is 'full Earth'; light reflected from the Earth lights the dark part of the Moon faintly. This is known as 'earthshine'.

Ecliptic The plane of the Earth's orbit around the Sun, seen projected against the stars. The ecliptic is the Sun's apparent annual path through the *zodiac* constellations.

Ejecta Fragmental material thrown out during the formation of a crater. The ejecta from volcanic explosions normally contains a high proportion of lava fragments. For craters formed by meteorite or missile impact, or artificial explosions, the ejecta consists of excavated bedrock.

Element An element is composed of atoms having identical atomic numbers and therefore identical chemical properties.

Equinox A point of intersection between the Sun's apparent annual path and the Earth's equatorial plane. When the Sun is at equinox (on 21 March and 23 September) it is overhead at the Earth's equator.

Faults A fracture in the Earth's crust, along which rocks on one side have been displaced relative to the rocks on the other.

Feldspars An important group of rock-forming minerals. They are all alumino-silicates with varying amounts of potassium, sodium, calcium and barium. The potassium and sodium feldspars (orthoclase and albite respectively) are the *alkali feldspars*. *Plagioclase* feldspars form a continuous series from sodium- to calcium-rich end members (see *Orthoclase* and *Plagioclase*).

Flow Fronts The scarp produced at the front of any flow that was sufficiently viscous at the time of formation not to thin out at its margins. The height of the flow front represents the thickness of the flow at that point.

Gibbous Phase A phase of the Moon or a planet when the disk of the body is more than half illuminated. The Moon is gibbous near first and last quarters.

Graben A trough of terrain that has been formed by down-faulting between two parallel faults.

Gutenberg Discontinuity The surface separating the mantle of the Earth from the core below; this lies at a depth of about 2,900 km below the surface.

Highlands (Moon) The relatively bright, densely cratered terrains on the Moon. These areas tend to be higher than the maria (see *Maria*).

Hornblende An important dark coloured rock-forming mineral of the *amphibole* group. It has a variable composition but is essentially a hydrated silicate of calcium, magnesium, iron, sodium and aluminium. It often forms needle-like crystals, and occurs in intermediate and acidic rocks. It is also common in metamorphic rocks.

Hydrosphere The regions of the Earth's surface covered by water as oceans.

Ignimbrite A word used with slightly different meanings in various parts of the world. In Britain it is used in the same sense as Ash Flow in American usage. The word means 'fiery rain cloud'.

Ions An electrically charged atom or group of atoms. Oppositely charged ions may be attracted to bond together, while similarly charged ions repel each other.

Ilmenite An ore mineral consisting of iron-titanium oxide $FeO.TiO_2$ which is the commonest industrial source of the metal titanium. Frequently found in small amounts in basic igneous rocks. Samples of igneous rock from the lunar Mare Tranquillitatis contain a high proportion of ilmenite.

Inertia This is a measure of the force needed to change the velocity of a moving body.

Insolation Energy received from the Sun in the form of radiation. On the Earth and Moon this amounts to 2 calories per square centimetre per minute.

Intrusive Rocks Those igneous rocks that solidified below the Earth's surface. They were formed by the intrusion of magma into pre-existing rocks where they formed magma bodies of a variety of sizes and shapes (see *Magma Chamber*).

Island Arcs Chains of islands forming arcs on the surface of the globe. These are sites of volcanic and earthquake activity. They are usually bordered by a deep oceanic trench on the convex side of the arc. Examples of island arcs include the Aleutian islands off Alaska, and the Kurile islands north of Japan.

Isochron A method of presenting rubidium-strontium data, in which analyses of samples from the same system are plotted together. In ideal cases the data points lie on a straight line whose slope is proportional to the age of the system.

Isotope Some *elements* are composed of *atoms* which, while still having identical atomic numbers and chemical properties, differ in mass. The term isotope refers to atoms of such an element, having a given mass, e.g. $strontium^{87}$.

Jovian Planets The giant planets Jupiter, Saturn, Uranus and Neptune. These planets have low densities and extensive atmospheres; investigations show the atmospheres to contain methane, hydrogen and ammonia.

Levee The raised embankment on the sides of a water channel or a lava channel.

Libration The apparent 'wobbling' of the Moon's disk during a lunar month.

Limb (of Moon) The edge of the visible disk of the Moon. Surface features at the limb are greatly distorted by foreshortening, when seen from the Earth.

Line of Nodes In the case of the Moon this is the line of intersection of the orbits of the Moon about the Earth, and the Earth about the Sun.

Lunation One lunar synodic month, or complete cycle of lunar phases (see *Synodic Month*).

Magma Molten rock; generated on the Earth by melting of the upper part of the mantle or lower crust. Once formed magma rises through the crust to give the intrusive rocks, or is erupted at the surface as lava.

Magma Chamber An underground reservoir of magma situated just below a volcano, or actually within the volcano. From this reservoir, lava is fed by dykes or up pipe-like conduits to craters from which it is erupted.

Magnetic Field The region surrounding a magnet, or magnetic object such as the Earth, over which its magnetism has influence.

Magnetometer An instrument for measuring the strength of magnets or magnetic fields.

Mare Ridge (*Wrinkle Ridge*) Ridges (formerly known as wrinkle ridges) on the lunar maria. These are composed of material having the same albedo as mare material. Often they have low angles of slope and are more readily identified under low lighting conditions. They have widths of up to a few hundred metres and heights of several tens of metres.

Maria (Singular: Mare) These are the dark level areas of the Moon (cf. *Highlands*). Almost all the extensive marial areas occur on the nearside of the Moon and consist of volcanic fillings to large basins. The basins are of two types: the circular maria presumed to have formed by excavation during the impact of asteroidal-sized bodies; and the irregular maria which are topographic depressions in the lunar surface. Such basins also occur on the far side but lack the dark fillings.

Mass A measure of the quantity of material (in grams, kilograms, etc). Mass should not be confused with weight which depends on the force of gravity. Thus an object of the same *mass* will weigh less on the Moon than it will on Earth because of the difference in the gravitational pull.

Mass Spectrometer Apparatus for determining the mass of an element. Different isotopes of an element may be distinguished and the isotopic ratios determined.

Mass Wastage Downhill movement, under gravity, of rock debris and soil. Small, slow movements give rise to soil-creep; larger movements give mudflows, etc.

Metamorphism The transformation of existing rocks into new types by the action of heat and/or pressure. This process usually takes place at depth in the roots of mountain chains or close to intrusive igneous bodies.

Meteorites Small cosmic bodies that have landed on the Earth's surface. These consist of metallic or stony material: the former being of nickel/iron alloy and the latter of silicate and oxide minerals, many of which are typical of terrestrial rocks. These bodies are called *meteoroids* in space, and *meteors* when they pass through the Earth's atmosphere as 'shooting-stars'. Micro-meteorites are similar to meteorites, but have diameters of less than 1 mm.

Mica An important rock-forming mineral found in a variety of different forms (see *Biotite* and *Muscovite*). The micas have a well-developed cleavage allowing them to be split into thin, elastic sheets and have good insulating properties.

Mohorovičic Discontinuity The surface separating the crust of the Earth above from the mantle below; it is at a depth of about 70 km below the surface of the continents, and 6 to 10 km below the floor of the oceans.

Moment of Inertia The moment of inertia of an object is a measure of the force needed to change its rate of rotation. It depends on the position of the object, and on the distribution of the mass of the object.

Muscovite A white mica consisting of a hydrated silicate of potassium and aluminium, $2H_2O.K_2O.3Al_2O_3.6SiO_3$. It occurs commonly in acidic igneous rocks and also some metamorphic rocks (see *Mica*).

Olivine An important rock-forming mineral consisting of magnesium, iron silicate, $2(Mg, Fe)O.2SiO_2$. Normally found in basaltic rocks. Clear green varieties are used as gemstones.

Orthoclase A member of the group of minerals known as *feldspars*, consisting of potassium, alumino-silicate $K_2O.Al_2O_3.6SiO$. It is an essential constituent of acidic igneous rocks such as granite (see *Feldspar*).

Peneplain A land surface worn down by erosion to give a flat or gently undulating plain.

Perigee The point in the orbit of a satellite when it is at its nearest to the Earth (see *Apogee*).

Plagioclase A continuous series of feldspars from a sodium-rich end member to a calcium-rich end member. Sodium and calcium are interchangeable within the lattice of the crystals, allowing replacement of one element by the other. They are found in almost all igneous rocks, the calcium ones being typical of basic rocks and the sodium equivalents typical of acidic rocks. Calcium-rich plagioclase occurs in mare rocks returned by Apollo missions.

Planetesimals Any small solid body in orbit around the Sun or a planet. Thus the smaller asteroids or particles in the rings of Saturn may be termed planetesimals.

Precession The rotation of an axis of rotation such as that of the Earth or the Moon.

Pyroxene Important rock-forming minerals that are usually silicates of magnesium and calcium. They are typical of basaltic rocks and occur in rocks returned from the mare regions of the Moon.

Quartz A mineral consisting of silica, SiO_2. Found in more acidic igneous rocks such as granite. Because it is chemically resistant to breakdown it is a normal constituent of sedimentary rocks formed by the breakdown of igneous rocks (e.g. sands and sandstone).

Radioactivity Radioactive isotopes have unstable structures and decay to more ordered forms. Energy in the form of radiation is released by this change.

Regolith A surface layer of fragmental rock formed by the breakdown of the underlying bedrock. It is analogous to soil, but unlike soil is not formed by organic breakdown and does not normally contain organic material as an essential constituent. The lunar regolith, which appears to cover the whole of the Moon's surface, is thought to have formed by smashing up of the bedrock by meteoritic impacts.

Retrograde Motion Most of the bodies in the solar system rotate and orbit in the same sense. For example, seen from above its north pole, the Earth rotates anti-clockwise. This is direct motion. Retrograde motion implies that the sense of motion is clockwise when seen from the north pole. Venus rotates in the opposite direction from that of the Earth and is said to have retrograde rotation.

Reversed Magnetism A state of the Earth's magnetic field in which the magnetic north pole lies at the position of the present magnetic south pole, and vice versa. Thus a compass needle would point south instead of north during a period of reversed magnetism.

Rilles These are clefts or 'valleys' on the lunar surface. Straight and arcuate rilles are apparently graben structures formed by the subsidence of a trough on parallel faults (see *Faults*). Sinuous rilles superficially resemble meandering river valleys; although some scientists have argued that they were formed by water erosion, it appears most likely that they represent lava channels in lava flows.

Royal Society The premier scientific society in Britain, founded in 1645; King Charles II was an active patron of the society. Its earliest members were wealthy amateurs with diverse interests. To be elected Fellow of the Royal Society is a great honour for a British scientist.

Seismograph An instrument for recording earthquake waves.

Shock Metamorphism Alteration of the constituent minerals of rocks by the passage through them of shock waves. These cause high temperatures and pressures for periods of just a few micro-seconds. In the most extreme cases the rock may be melted, or new minerals formed such as diamond and coesite (a high-pressure variety of quartz). The internal structures of minerals become deformed in other cases.

Sidereal Month The time taken for the Moon to orbit the Earth once with respect to the stars: 27.3 days. This is also the time taken for the Moon to rotate once about its axis.

Silicate Minerals Minerals rich in silicon and oxygen. Silicon forms a strong chemical bond with oxygen, giving rise to a negatively charged group of atoms in which one silica atom is surrounded by four oxygen atoms: SiO_4. Sometimes these charged 'silicate groups' are linked to form chains or sheets, and combine with elements such as sodium, potassium, magnesium, calcium, iron or aluminium. Silicate minerals are the commonest in the Earth's crust and mantle; also the Moon rocks so far returned.

Solenoid A coil of wire wound round a tube. When direct current is passed through the wire it becomes magnetic, so that an iron rod is drawn into the tube. A solenoid is a common part of the starting system of a motor-car.

Solstice There are two solstices in the Sun's annual path (seen from the Earth), one when it is apparently over the tropic of Cancer (22 June) and the other over the tropic of Capricorn (22 December). These represent the northern and southern limits of the Sun's motion and consequently they mark midsummer in the northern and southern hemispheres respectively.

Specular Reflection Light striking a surface is usually scattered and reflected in all directions; but if the surface is flat, the light is reflected in one direction: this is the direction of specular reflection. A mirror reflects light specularly.

Stadia (Singular: Stadium) Measure of distance in Greek Antiquity. Probably equals about 185 metres.

Stellar Magnitude A relative measure of the brightness of stars (and by extension, planets and other astronomical bodies). A magnitude difference of 5 corresponds to a factor of 100 in brightness. The brightest star, Sirius, is of magnitude -1.4, and the faintest naked-eye star of magnitude $+6$. The magnitude of full Moon is -13.3.

Synodic Month The average time interval between successive new Moons: equal to about 29.5 days.

Terminator The line on a planet or the Moon dividing the bright (day) and dark (night) hemispheres. Near the terminator on the bright side the Sun's rays strike obliquely, giving long shadows and picking out topography clearly.

Turbidity Currents These form underwater and consist of a dense and turbid mixture of rock particles and water. These may move at high-speed on the sea floor, and have been known to break underwater telephone cables.

Unconformity A surface of erosion or non-deposition in a succession of rocks. Often the older strata, below the unconformity, dip at a different angle from the younger rocks above.

Vesicles Small cavities commonly found in lavas. They are normally caused by expanding gases in the lava. Lunar lavas have been found to be particularly vesicular in character. Large irregular vesicles are sometimes termed 'vugs'.

Zodiac The band of constellations, stretching round the heavens, that the Sun moves through during the course of the year. The Moon and most of the planets also move through the Zodiac.

Index

Abyssal plains, 84–7
ACIC (Astronautical Chart and Information Center), 25
　Moon charts, 12–14
Agriculture, space view of, 98
Albedo
　Earth, 89–90, 97
　Moon, 89
Alphonsus, 134
Anatomy of Earth (Robinson), 37
Andesitic rock, 55–7
Anaximander's theory proved 2,500 years later, 99
Anorthositic lunar rock, 131
Apollo landings, knowledge from, 26, 32–3, 35–6
Aristarchus, orbital theorist, 10
　crater, 101, 104, 138–9
　plateau, 101
Aristotle, 10–11
Asthenosphere, 46–7
Astronaut, dog as first, 7
Astronomy, 6,000-year evidence of, 9
Atmosphere
　pollution of, 87
　scattering properties of Earth's, 91, 95
Aubrey Holes of Stonehenge, 9
Australia, ice age evidence in, 73

Barycentre, Earth-Moon, 19–20
Basaltic rock, 55–7, 131, 152
Bauxite product of weathering, 68
Beard, D. P., 13
Benioff seismic zones, 44–6
Biotite, 151–2
Blake Plateau, 85
Blocking temperature, 151
Blue Earth, 91
Bombardment, meteorite, 110–11, 120, 134–5
Boneff, W., 13
Borman, F., astronaut, 7
Brahe, Tycho, 10
Breccias, lunar, 135, 138

Calderas, volcanic, 63–4, 134
Calendar, moonrise basis of megalithic, 9
Cassini, Moon mapper, 11, 12
Chaldeans, astronomy and, 9
Clouds pictured from space, 88, 91, 93–4, 96–7
Cones joined by fissures, 134
Continental
　deposits, 78–80
　drift, 8, 39, 154
　shelves, 38, 85
　slopes, 85
Copernicus crater, 103, 138
Coral reefs, 87
Coriolis forces, 93
Craters, volcanic, 63–4
Crop health, space pictures show, 98
Cuvier, George, 15
Cyclone warnings from space, 94

Darwin, Charles, 16, 87
Da Vinci, Leonardo, 12, 15, 41
Decay rates, identical, 150
Deltas, 78, 81–2
Deposits, 78–85
　biochemical-chemical, 82, 84
　clastic, 82–4
Deserts, erosion in, 76–8
Drift, 78
Dykes, 16, 49

Earth
　crust movement cycle, 46–7
　density, 28–9, 32
　interior structure studies, 29–36
　internal heat, 34
　liquid magnetic core, 33–4
　magnetic fields, 33–4, 39–41
　　anomalies, 33, 39, 41
　-Moon physical comparisons, 17
　Moon view of, 8
　orbital path, 19–20
　original theories, 28, 36
　science-based theories of, 15
　sole single-satellite planet, 17, 27–8
　space programmes spur study of, 13, 16
　space view of, 8
　surface hardly scratched, 29
　surface in continual change, 8, 37–47, 66–87
　theology inhibits knowledge of, 15
　tidal distortions of solid bulk, 18
　unique atmosphere of, 90
　volcanic clues to mantle of, 33
　water's role on, 8, 66–87
Earthquakes, 29–32, 37–9, 43–6
　areas of most intense, 44
Earthshine, 20, 89
Eclipse, -s, 9–10, 20–1, 95, 109
　annular, 21
　temperature measurements of Moon during, 21, 109
　umbra, 21
Ecliptic path, 20
Egg-shaped Moon, 36
Egyptians, astronomy and, 9
Ejecta blanket, 138
El Campo, 13
Equatorial bulge, Earth's, 22, 36
Eratosthenes measured Earth by shadows, 10
Erosion, 66–77
　agricultural, 87
　vegetation absence and rapid, 77
　weight loss starts surface elevation, 47
Everest
　marine fossils on, 41
　still growing, 41–2

Fault, -s, faulting, 8, 39, 45–6, 64, 82
Fissure volcanoes, lunar, 134
Fluviatile sediments, 78
Fold, -s, folding, 16, 42, 44, 82–3, 139
Fossils as rock daters, 16
Fragmental rock from Moon, 131

Gagarin, Yuri, 7
Galileo, 11

Geochronology, 149–56
Geology
   artificial influences by man on, 87
   unlimited process time, 16
Geometric measurements pioneering, 10
Geosyncline, 42–4
Gilbert, G. K., 120
Glacial
   deposits, 78
   formation, 71–3
Glass beads
   cratered, 120
   from regolith shock, 138
Glass shards in volcanic ash, 60
Glenn, Major, 7
Golf on the Moon, 105
Graben structures, 134
Granitic rock, 16, 55–9, 131
Gravitational field irregularities, 88
Great Barrier Reef, 87
Great Glen of Scotland fault, 46
Gruithuisen, F. P., 13
Gutenberg discontinuity, 32, 34
Guyots, 85–7

Harbour vanishes overnight, 37
Harriot, Thomas, 10–11
Herschel, Sir John, 14
Herschel, Sir William, 10, 13
Hooke, Robert, 11
Horneblende, 68, 151–2
Hurricane warnings from space, 94
Hutton, James, 15–16

Ice, geological effect of, 71–3
Igneous intrusions, metamorphosis adjacent rocks, 16
Igneous rocks, 48–64, 131
   chemical and mineralogical composition of, 54–5
Ignimbrites, 59–64
Illustrations of the Huttonian Theory of the Earth (Playfair), 15
Ilmenite, lunar, 131
Imbrium basin, 121
India, ice age evidence in, 73, 83
Infra-red
   Earth emission variations, 88
   space photography less obscured, 94–5
Insolation, 93, 108
Intrusive rocks, 16
Island arcs, 85
Isochron plotting, 153
Isostatic adjustments, 47
Isotope decay, 149

Karst topography, 71
Kelvin, Lord, 34
Kepler crater, 104
Kepler, Johannes, 10
Kimberlites, 33
Kirwan, R., 15
Krieger, lunar cartographer, 14

Laki, Iceland, eruption at, 55
Landforms, change through time of, 16, 76
Langren, M. F. van, 11
Langrenus, 11
Large ray craters, 138–9
Lava remelted becomes rock, 135
Leonidas, 37
Libration, 13, 19, 22–5
   aids limb measurement, 25
   Galileo discovers Moon, 11
Limb, lunar, 24–5, 100
Line of nodes, 20
Linné Crater, 13
Lippersheim, Hans, 10
Lithosphere, 46–7
Locke, R. A., 14
Lohrmann, T., 13
Lovell, astronaut, 7
Lucian of Samos, 2nd Century SF writer, 10
Lyell, Charles, 38

Madler, J. H., 13
Magma, -s, 48–64, 151, 153–4
   catastrophic eruptions from siliceous, 59–63
   dome for volcanic pipe, 59
   effect of water on, 51–4
Magnetic field, 33–4, 39–41
   a radiation shield, 33
Mare Crisium, 120
Mare Fecunditatis, 8
Mare Imbrium, 26, 106, 120, 130
Mare Serenitatis, 106
Mare Tranquillitatis, 130–1, 155
Marginal plateau, 85
Marine deposits, 82–5
Marine fossils on mountains, 41
Mars, 28, 90–1
   non-spherical satellite of, 18
Mascons, 26, 105–7
Matthews, D. H., 39, 41
Mayer, Tobias, 12
Megalithic knowledge of astronomy, 9
Mellan Claude, 11
Mesospheric processes, global, 88
Metamorphism, geological, 16
Meteor Crater, Arizona, 138

Meteorites, 11, 32–3, 120, 134
   dating, 154–5
   iron, 33
   meteors and, lunar, 8, 110–11, 134
   silicate mineral, 33
Micrographia (Hooke), 11
Micrometeorites, 8, 110–11
   crater rock fragments, 138
Military aspect of space pictures, 94
Minerals, Moon cf. Earth, 131
Mobile belts, 37–9, 41, 44, 46
   map of, 39
Moho (Mohorovicic) discontinuity, 32, 38, 43
Mond Atlas (Krieger), 14
Mont Pelée eruption, 59–63
Moon
   abandoned theory of life on, 13
   age fixed by surface rock, 26, 139, 155–6
   atmosphere, 108
   axial rotation, 21–2
   changeable surface heights of, 18, 25–6
   colour clues to history, 105
   craters,
      age softens rims of, 103, 134–5
      believed volcanic, 11
      terraced interiors of, 134
      theories about, 13–14, Chap. 9
   density, 28–9
   egg-shaped, 36
   far side heavily cratered highlands, 121
   first map of, 11
   gravity, 105–7
   highlands, 24, 35, 105–6, 120–1, 139
   hoax story of life on, 14–15
   ice possible underground, 110
   interior non-spherical, 36
   interior structure studies, 29, 32, 34–6
   internal heat, 35
   lava, 8, 130–4, 139
   lowlands areas mistaken for sea, 12–13
   magnetic field, 33–4
   Man explores, 7–8
   mare areas, 36
   mare ridges, 100, 130
   maria, 24, 27–8, 35–6, 104–7, 120–1, 130–4
   marial basins, 121
   measuring shape of, 24–6
   meteorites and micrometeorites, 8, 110–11, 120, 134–5, 138

minerals earthlike, 131
no life on, 8
non-sphericity of, 25, 36
observations
    from Earth, 100
    from space, 100–3
    sunlight angle affects, 103–4
once active, 156
orbital path, 19–20
origin theories, 26–8, 36
phases, 20
photographs from unmanned spacecraft, 7
proton bombardment, 111
regolith, 35, 108–10, 134–5, 155
rocket gas contamination, 108
rocks, study of, 8
seismograph traces different from Earth, 32
solid core assumption, 34
space programmes spur study of, 13
sphericity measurements, 24–5, 36
surface
    bombardment, 110–11, 120, 134–5, 138
    colour, 104
    first unmanned landings, 7
    height differences, 25–6
    temperature, 108–10
        during eclipse, 109
    terrain, two distinct types of, 120–1
volcanic evidence, 131–4, 139
Moonquakes
    Apollo missions enable studies of, 32, 35
    regolith creep from, 138
Mountain building, 8, 41–4, 46
Muscovite, 68, 151–2

Neolithic knowledge of Sun and Moon, 9
New York Sun, 14–15
Nuclear craters, 139
Nuées ardentes, 63–4, 139

Observation as study basis, 16
Occultation, 108
Oceanic
    morphology, 85–7
    pollution, 87
    ridges, 85
    rises, 85
Oceanus Procellarum, 120, 130–1, 155

Olivine, 56–7, 68
    lunar, 131
Orientale basin, 121

Peneplain, 70
Permafrost, 110
Philolaus, orbit theory pioneer, 9–10
Phobos, non-spherical satellite, 18
Photogeology, 7
Photogrammetry
    to study inaccessible regions, 98–9
    problem of lunar, 25
Photographs
    aid Moon mapping, 13
    military aspects of orbital, 94
    Moon cf. Earth, 88–9
Plagioclase, lunar, 131
Planetesimals theory, 28
Planets, three influences on shape, 18
Plants as rock eroders, 68
Plate tectonics, 8, 46–7
Playfair, John, 15
Port Royal Harbour, Jamaica, 37
Potassium-argon dating, 151–2, 155
Precession, 29
Principles of Geology (Lyell), 38
Prinz crater, 14
Protons attack Moon surface, 111
Pyramids sited astronomically, 9
Pyroxene, lunar, 131
Pythagoras, 9

Quakes, Moon cf. Earth, 32

Radioactive
    decay
        aids rock age fixing, 149
        heat source, 34, 36, 48
        elements on Moon, 35
Red Mars, 91
Reversed magnetism, 33, 41
Riccioli, G. B., 11
Rivers
    erosion transporters, 70, 78–82
    space pictures of, 97–8
Robinson, T., 37
Rock
    ages, 149–56
    content tables: Earth cf. Moon, 57, 131
    crumpling, 42–4
Rotation
    distorts sphericity, 17–18
    space pictures show cloud, 93
Rubidium-strontium dating, 152–5
Russell, John, 12
Rutherford, Ernest, 149

San Andreas fault, 45–6
San Francisco earthquake, 45–6
Satellites, non-sphericity of, 18
Schmidt, J. F. J., 13
Schröter, J. H., 13
Scientific geology, pioneer of, 15
Sea
    erosion by, 73–6
    level variations, 73, 76, 81–2
    -mounts, 85
    recession, 37
Secondary impact craters, 138
Sedimentary rocks, mountains from, 42
Sedimentation starts rock formation, 77
Seismology, 29–32, 44
Shelf break, 85
Shelves, continental, 38, 85
Shepard, Alan, 105 n.
Shock metamorphosed material in Moon rock, 138
Shortening, 42–4, 46
Silicate rock minerals in ranged stability, 68
Sinus Medii, 11
SiO_4 tetrahedron, 54
Smith, William, 16
Solar wind, 33, 108, 110–11
Solifluction, 70
Southern Africa, ice age evidence in, 73
Space pictures aid Earth knowledge, 88–99
Space programmes, knowledge from, 13, 16, 18, 26
Specular reflection, 89
Sphericity erodes mountain heights, 18
Sputnik 1, first orbiting spacecraft, 7
Stable areas, 37, 44
Stonehenge, Sun and Moon positions and, 9
Sub-surface water, rocks eroded by, 71
Surface
    continuous changes in, 8, 37–47, 66–87
    changes, causal forces, 66
    tidal distortions of solid spheres, 18
Synodic lunar month, 20

Telescope, invention of, 10
Thales, astronomer, 9
Theory of the Earth with Proofs and Illustrations (Hutton), 15

Thermopylae, Pass of, 37
Thorium-lead dating, 154–5
Tides, distortions from, 17–18
Timekeeping, Moon for, 9
Time, unlimited geological processes, 16
Transcurrent faults, 46
Troughs, sphericity and, 18
Turbidity currents, 84–5
Tycho crater, 103, 109, 131, 138–9

Unconformities, geological, 16
Uniformitarianism, 15–16
Uranium-lead dating, 154–5
Usher, Archbishop, 15

Varved sediments, 78
Villages now below North Sea, 37
Vine, F. J., 39, 41
Volcanic
    theory of Moon craters, 120
    chemistry, 48–64
    lava from Moon, 8, 130–4, 139
Volcano, -es, 11, 13, 33–4, 37-8, Chap. 5
    escaping gases as safety valve, 54
Water
    cyclical functions of, 66
    Moon surface has no, 110
    underground, 71

Weathering cycle, 66–70, 77
Weather patterns, space pictures emphasise, 93–4
Wegener, Alfred, 39–41
Weisberger, Herr, 13
Werner, A. G., 15
Wilson, J. T., 46
Wind erosion in deserts, 76–8
Wren, Sir Christopher, 11

Zodiac constellations, 20, 22